Patrick Moore's Practical Astronomy Series

GW00469153

For other titles published in the series, go to
www.springer.com/series/3192

Deep-Sky Video
Astronomy

Steve Massey and Steve Quirk

Springer

Steve Massey
Hervey Bay, QLD
Australia

Steve Quirk
Mudgee, NSW
Australia

ISBN 978-0-387-87611-5   e-ISBN 978-0-387-87612-2
DOI: 10.1007/978-0-387-87612-2

Library of Congress Control Number: 2008940645

Printed on acid-free paper

springer.com

Acknowledgements

The true rise of video as a popular and affordable, practical tool for imaging the night sky stems back to the mid-1990s, when small pockets of enthusiastic individuals around the world shared their experiences and knowledge mostly via the still early days of the Internet. Like most things in astronomy, the evolution of the technology has been spurred on by the enthusiasm and support of others combined with thousands of hours under star-filled night skies experimenting, measuring, and reporting. Either directly or indirectly, long chats by phone, at star parties, club meetings, or via e-mail, the production of this book would not have been possible without their help. First and foremost we would acknowledge the unceasing support of our life-long partners Sandra Massey and Janet Quirk. Along with friends and members of various discussion groups, such as the GSTAR-Users group, VideoAstro, and QCUAIG, our expressions of deep gratitude to Chris Wakeman, Darrin Nitschke, Rob McNaught, Gordon Garradd, Michael Clark, Allan Gould, Phillip Hinkler, Jim Ferriera, Ron Dantowitz, Albert Van Donkelaar, Dave Larkin, Mark Garrett, Jon Bradshaw, Robert Knox, Lindsay Sessions, Mike Holliday, John Sarkissian, Chris Toohey, the team at Australian Sky and Telescope Magazine, Sky Publishing USA, Tasco Australia, Steve Kerr, Dave Gault, Steve Wainwright, Dr. Andre Phillips, and Bob Balfour. Also, special thanks to Steve Chapman and John Vetter for testing our procedures for the book. We would also express our appreciation to John Watson, Maury Solomon, Peter Pushpanathan and Turpana Molina of Springer Publishers for their helpfulness, enthusiasm, and support of the project.

About the Authors

Steve Massey

Starting his journey in astronomy at the age of nine from the backyard of his family home in the early 1970s, Steve has pursued his interest on both amateur and professional levels.

In the mid-1990s he worked at the Anglo-Australian Telescope at Siding Spring Observatory and has frequently used a variety of professional telescopes (up to 1 m-class) to conduct video planetary imaging projects.

With over a hundred radio interviews and several appearances on TV, he is among a handful of pioneers that helped to popularize the use of CCD video cameras for use in astronomy and was an invited contributor on the subject for the *Oxford Astronomy Encyclopedia*.

Having numerous articles and images published in newspapers, video-camera periodicals, and astronomy magazines around the world he is also an author and co-author of six other top-selling titles, including *Video Astronomy, The Night Sky (A Guide to Observing the Solar System), Exploring the Moon, How Does the Night Sky Work, Space-Stars and Planets*, and *Atlas of the Southern Night Sky*. He is also co-author and contributor to published scientific papers concerning the planet Mercury.

Today, he operates a successful optical supply business in Queensland Australia.

The asteroid 14420 Massey (1991 SM) was named after him.

Steve Quirk

An amateur astronomer for nearly 30 years, Steve Quirk forged his reputation for excellence in astrophotography during the 1980s and 1990s, producing many inspiring photographs of the night sky, some of which have been used in magazines and books in Australia and overseas.

Not satisfied with obtaining images of only the most famous celestial bodies, he has built up an impressive photo library of many difficult and less-frequently observed objects, using self-taught skills in star hopping and the use of setting circles. He currently uses modern CCD-based video cameras to illustrate many of the fine deep-sky objects featured in this book. Many of his images are considered among the best in the world using integrated video technology.

He is a member of the Central West Astronomical Society (Australia) and has been a long-time assistant in a video meteor surveillance program run by Robert McNaught at Siding Spring Observatory.

Quirk was the winner of the 1988 Skywatch Bicentennial Astrophotographic Competition and received recognition at the Central West Astronomical Society's AstroFest David Malin Awards in 2005 for a 20-year-long sequence of images showing the movement of the star Proxima Centauri.

He has written several articles on astrophotography for (Australian) magazines and is co-author of *Atlas of the Southern Night Sky* with Steve Massey.

Today he spends most of his time imaging with video from his observatory in central west NSW Australia.

The asteroid 18376 Quirk (1991 SQ) was named after him.

Contents

Introduction

Deep-Sky Video Astronomy

Welcome to the wonderful world of video deep-sky imaging!

For hundreds of years after the first telescopes were pointed skyward, our only record of the celestial realm was limited to what our eyes could see. With careful interpretation observers shared their views in sketches or paintings.

When chemical-based plate and film photographic tools became available, things began to change rapidly in the world of astronomy, but it was not until the development of greatly improved sensitivity to faint light that astronomers could efficiently record those well-traveled photons from distant stars, nebula, and galaxies. This truly opened the doors to wider scientific analysis and understanding of the universe around us.

In the decades that followed, developments in practical electronics soon ushered in a new picture medium – television. The enormous studio cameras that once beamed pictures into our homes were, like television, also based on a cathode-ray vacuum tube (CRT) design. CRT televisions and computer monitors are still used today.

As the rising bell curve of technological improvements continued, electronic components were being dramatically miniaturized, and the valves in our radios and televisions were soon replaced with small silicon semiconductor transistors. In a few short years, engineers figured out how to make them even smaller allowing them to pack thousands of these transistors together into tiny arrays called an integrated circuit (IC), and the possibilities were endless. In 1969, Bell Laboratories in the United States developed a light-sensitive array called a CCD, which is an acronym

Fig. 1. The "Trifid Nebula" (Messier 20) in Sagittarius. A fine example of what modern video technology can produce.

for charge-coupled device, and combined with powerful modern microcomputing; this amazing invention revolutionized astronomical imaging on many levels.

In general consumer goods, the CCD has rendered almost completely redundant the use of film-based cameras for taking snapshots and making movies. CCD technology lies at the heart of all professional and domestic camcorders, pocket cameras, scientific imagers, single-lens reflex (SLR) cameras, and webcams. Even cell phones take pictures and movies!

Today, the backyard astronomer can not only produce amazing images but also yield scientifically useful data with compact, lightweight, low light-sensitive cameras. Moreover, taking a great portrait of the night sky is no longer limited to the once elite amateur astrophotographer. But CCD imaging still has its elite few who well understand the rules for getting the most out of a camera and how to extract more from the final picture using advanced image manipulation software. Video imaging is no different.

Like the old film movie camera, a modern CCD video camera simply takes a rapid sequence of short, individual exposures and when viewed live on the screen or in playback mode, gives the illusion of motion. The benefits of multiple rapid exposures has long been known to astronomers as a useful means of capturing random moments of exceptional atmospheric seeing in the constantly moving air mass that surrounds our planet. By selectively plucking out individual exposures, astronomers have been able to produce diffraction limited portraits of the moon and planets to create images with amazingly sharp detail.

Delineating the differences between "conventional" analog cameras and those that can only function by computer interface (a webcam for example) has become somewhat clouded over the years and some incorrectly refer to closed-circuit television (CCTV) cameras as a webcam. Indeed, there have been recent developments whereby the best of both technologies now exists. But since both systems produce a stream of rapid exposure pictures that can be recorded as a movie file, we prefer to call them image streaming cameras (ISC). At the end of the day, whether the source is a cell phone, camcorder, webcam, or other ISC, the moving pictures they create is referred to as video and its use for imaging the celestial wonders aptly called video astronomy.

But can these short-exposure cameras be used to reveal dim objects such as galaxies and nebulae? In the past, when even the most-sensitive camera used for general surveillance work in low-light situations was applied to a telescope, the answer was simply "No," unless it was combined with an inordinately expensive, military-grade image intensifier.

Some adventurous amateurs sought a more affordable approach and turned to modifying basic webcams to perform way beyond their intended design. With a little circuit tampering to enable extended exposure times combined with some basic thermal cooling, they produced quite good results. When used in conjunction with creatively designed software such as COAA's AstroVideo, which co-adds a predefined number of images output by a camera on the fly, the result is a final image revealing vastly amplified detail.

However, today it is indeed a reality that we can now view deep sky wonders in virtual real-time as well as create the equivalent of a long-exposure photograph using off-the-shelf frame-accumulating video cameras. With the ability to internally accumulate tens or hundreds of short exposures, these cameras make viewing and photography of star clusters, nebulae, and galaxies (even from light-polluted suburbs) a breeze. Like any other CCD camera image, postprocessing at the computer is the ultimate key to producing a true, aesthetically pleasing result, and many images, when taken from near-city regions, often surpass film work done from dark country skies.

Sometimes referred to as integrating video cameras, whether they are analog or digital output, these frame-accumulating video devices are available from several specialist astronomy product vendors worldwide. Supplied with various modifications, supporting software, or other features they are packaged specifically for use in astronomy. Frame-accumulation cameras are the perfect tool for sharing the wonders of the night sky with the public, friends, and family. No need for climbing stepladders to peer into the eyepiece of a giant Newtonian, and no need to refocus an object for each individual. Viewed on a TV monitor they offer outstanding contrast with ultimate eye relief that it is second to none.

As sad as it may sound to purist visual observers, stories of star parties where visitors have been seen cueing to view a deep-sky object on a video monitor, is becoming more commonplace, especially in the case of super aperture scopes 16 inches and larger. Deep-sky capable video cameras breathe new life back into those old, unused, small telescopes that have been relegated to the back shed and collecting dust. These specialized video cameras can now enable views of fainter objects normally only within the visual limits of much larger aperture instruments.

We have personally experienced the excitement and utter amazement of seeing deep-sky objects in real time through our telescopes that were either on or beyond the visual limit for that scope, even with fully dark-adapted eyes. Every time a search is made for those faint new objects, they are found with relative ease and a great deal of satisfaction. The stars that can be seen live on a monitor are around 1–1.5 magnitudes deeper than the best visual limit for a given telescope, effectively doubling the aperture. So, for an extremely cheap telescope upgrade, it is simple, get your hands on a deep-sky capable video camera!

Our goal in this practical guide is to explain the essentials about viewing and imaging with frame-accumulation cameras. From the main camera features to setting up for use at the telescope, we cover the image-capture process with practical steps and the most important basic post-processing techniques you can use to create wonderful deep-sky portraits.

Deep-sky videography is a fun, convenient, and often very practical low-cost imaging alternative with a diverse range of applications in amateur and professional astronomy. It is our sincere hope that we will pass onto you, through this book, the amazing benefits and knowledge of our experiences and others to ensure you gain the most from this wonderful and simple-to-use imaging medium.

Fig. 2. An image of videographers at work and play, illuminated only by star and moonlight, recorded with a GSTAR-EX video camera and 135-mm lens. Courtesy of Darrin Nitschke.

CHAPTER ONE

Using Video for Astronomy

Video cameras today are being used for a wide and diverse range of activities in astronomy. A few are listed here.

- The imaging of planets, moons, asteroids, and comets
- Occultation imaging and timings
- Solar imaging
- Meteor and fireball patrols
- All-sky monitoring for observatories
- Supernova searching and confirmation
- Deep-sky imaging

The most alluring features of video are the lightning-fast exposures and long recording times. In fact, fast-exposure, image streaming cameras make it possible to record detail on the space shuttle and the International Space Station (ISS) as they pass quickly overhead. Today, specialized video cameras have the ability to record very faint objects, and with some post-processing at the computer, the results can often rival more expensive single-image, cooled astronomy cameras (Fig. 1.1).

But before we leap head first into the practical side of things, you may find it interesting and perhaps beneficial to understand a little of how a CCD works and its application in a video camera to produce the images we want to obtain with a telescope.

S. Massey and S. Quirk, *Deep-Sky Video Astronomy*,
DOI: 10.1007/978-0-387-87612-2_1, © Springer Science + Business Media, LLC 2009

Fig. 1.1. An example of the high-speed exposures possible with video cameras and their ability to record random events.

1.1 The Image Sensor

The heart of the modern camera is its image sensor, or light-sensitive detector. Called a CCD, this solid state analog device consists of an oxide silicon substrate and an array or a matrix of vertical and horizontal light-sensitive electrodes (detectors) on the surface called pixels (short for *picture elements*). The size of each pixel is measured in microns (μm) and varies from one manufacturer's design to another. In most quality video cameras these days, pixels typically range from 6 to 10 μm, and the array is protected from undesirable elements, such as moisture and dust, by a thin glass window.

Light falling onto each pixel is converted proportionately into an electronic charge, and a combined pattern of the exposed pixels is read via electronic encoders and decoders, forming a digital interpretation of the image. The image produced is comprised of a discrete range of numerically defined values of light intensity or shades of gray from pure black to white, which our eyes perceive on a video monitor as different amounts of shadow and light.

Important to astronomy have been the continued developments in low-light sensitivity performance. Companies such as Sony Corporation developed improved CCDs that can be used in virtual "no-light" situations to meet the high demand of the growing security surveillance industry. One innovation, which is found in Sony's ExView HAD series, is the placement of microscopic lenses over each pixel, thereby more efficiently focusing photons on each detector across the entire array. This CCD boasts extremely low dark current qualities.

Depending on design for a given purpose, not all CCDs have the same inherent characteristics. Some are manufactured to respond more or less efficiently over a specific spectral range. For most domestic cameras, this is centered typically around the 550 nm yellow part of the visual spectrum, which replicates the strongest area of responsiveness we can detect with our eyes. The cameras generally range in spectral sensitivity from 350 nm to 1,100 nm but are often rated as 400–1,000 nm. Being quite sensitive in the infrared, it is also considered very beneficial in some astronomical applications (Figs. 1.2 and 1.3).

An image sensor efficiency rating for the detection and conversion of photons is known as detector quantum efficiency. The higher the rating, the better the sensor is for deep-sky imaging. A rating of 100% is considered a perfect detector, but most of the highly sensitive and affordable video cameras we find today have typical quantum efficiency ratings of around 50% or slightly better. Sony's ExView HAD series, used in many deep-sky video cameras today, boasts a quantum efficiency improvement of 18–20% above what is typical.

Fig. 1.2. At the heart of all modern digital cameras, the *CCD* collects photons and converts them into an electrical charge to create an image.

Spectral Sensitivity Characteristics

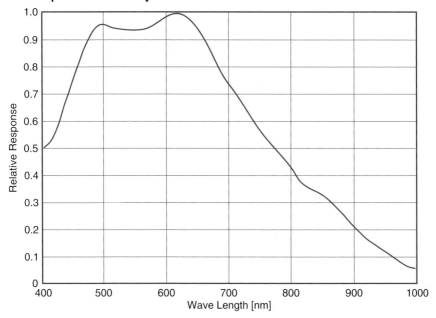

Fig. 1.3. This graph represents the spectral response of an unfiltered *CCD* image sensor found in most commercial cameras.

1.2 Efficiency in Astrophotography

Like the long exposures needed to record deep-sky objects with film, CCD cameras are also capable of storing or integrating a single image for extended periods but can achieve similar, if not far better, results in only a fraction of the time. Also, because we live on a constantly turning world, long-exposure astrophotography requires a well-aligned mount and accurate guiding to achieve those aesthetically desired round or pinpoint stars. Depending on the optical magnification, alignment, and tracking accuracy, short-duration exposures, like a subsecond snapshot, are far less demanding.

Whether producing digital USB, Firewire, or an analog signal output, webcams, camcorders, or CCTV-specific cameras are inherently short-exposure devices that pump out "still images" in rapid succession. Viewed on a monitor or recorded and played back, this series of fast-changing pictures give the illusion of real-time motion.

Short exposures mean less time for photons to build up a significant enough charge in the CCD for it to reveal faint structure in deep-sky objects. However, since the advent of frame accumulation short-exposure cameras, even the smallest amount of charge detected by a highly sensitive CCD, can be revealed as though the exposure time were much longer. This is achieved by electronically co-adding numerous exposures using an internal image memory buffer.

There are essentially two different video picture output techniques used by a CCD to read out the image we eventually see on a monitor or display. These techniques are known as progressive scan and interline transfer. The latter produces an interlaced TV picture (Fig. 1.4).

1.3 Progressive Scan

Webcams and all modern consumer digital camcorders generally use progressive scan technology, which outputs the image recorded across the entire active pixel array at once. This technique is very effective in situations of high-speed motion, offering less smeared recordings of things like a car race or in astronomy, for freezing out the sharpest moments from rapidly changing optical distortions caused by air turbulence in our atmosphere. Progressive scan cameras are particularly useful for high magnification work, such as lunar and planetary imaging.

1.4 Interlaced TV Images

The foundation of the interlaced picture at basic TV standard dates back to the days of early TV and still has a governing influence over much of the new digital standards, for reasons including backward compatibility for digital conversion and aspect ratios. The picture we see on a conventional 4:3 aspect TV monitor, for example,

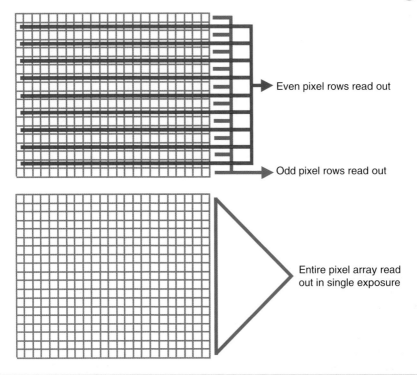

Even pixel rows read out

Odd pixel rows read out

Entire pixel array read out in single exposure

Fig. 1.4. Two *CCD* readout formats. *Above,* interline transfer CCD and *below,* a progressive scan CCD.

is produced by an electronic gun attached to the back of a phosphorous-coated vacuum-tube display (also known as a cathode ray tube, or CRT), which is illuminated when struck by electrons. As it traces out an image from the top of the screen to the bottom, phosphors illuminate with varying levels of brightness, according to the intensity of the electron beam.

Each line traced out across the screen takes around 64 microseconds (μs) to complete and is called a scan line. The duration between successive scans while the gun is off and returning to a point just below the starting point of the previous scan is known as horizontal line blanking time. A full TV picture (a single frame of video) is made up of two interlaced sets of these completed traces from top to bottom, called odd and even scan line fields.

In the case of the phase alternating line (PAL) television standard, the odd scan lines are traced from left to right and top to bottom of the CRT in about 1/50 of a second (20 ms), creating the odd field, which is comprised of 312.5 lines. The off time for the electron gun to move to the top of the screen again is the vertical blanking period, after which a series of even scan lines are then sprayed to fill in the blank even rows of the picture from top to bottom. This occurs within the next 1/50 of a second and makes up the even field. Each field, therefore, represents only half of the complete picture (Table 1.1 and Fig. 1.5).

Table 1.1. Basic PAL and NTSC characteristics[a].

TV system	PAL 625/50	NTSC 525/59.94
Scan lines per video frame	625	525
Video frame display rate/s	25	30 (29.97)
Scan lines per field	312.5	262.5
Field frequency (Hz)	50	60 (59.94)
Field display time (ms)	20	16.68
Active lines (containing image) per field	287.5	242.5
Total active lines per video frame	575 (576)	475 (476)
Line trace time (µs)	64	63.55
Vertical blanking lines	25	20
Horizontal line blanking time (µs)	12	~10
Active (picture display) line trace time (µs)	52	52.65

[a]Some numbers are rounded off and serve only as a rough guide

PAL phase alternating line, *NTSC* National Television Systems Committee

Fig. 1.5. Interlacing of odd and even scan lines on a conventional *CRT* television display.

The combination of these two fields (interlaced) produces the entire picture (a single frame of video) comprised of 625 scan lines displayed every 1/25 of a second (40 ms). The actual picture information is contained within the bulk of these scan lines (the active lines), but some of the scan lines are reserved for delimiting frame boundaries, synchronization timing, and other PAL or National Television Systems Committee (NTSC) signal carrier requirements and are cropped from the visual image displayed. Furthermore, it is interesting to note that only half of the first and last active scan lines contain picture information. For convenience purposes these lines are essentially combined into the active picture line count as making up an additional scan line, that is, $287.5 + 287.5 = 575$ scan lines. When doing number rounding for simplicity in digital sampling, however, the blank part of these scan lines is taken into account, thus being considered as an additional scan line - hence, the common 576 we encounter.

Interline transfer CCD video cameras producing an encoded carrier-based analog signal output also process two sets of line-matching exposures (interlaced fields) to meet this convention by reading out successive rows of pixel data just like the alternating scan lines producing the odd and even interlaced fields on a TV monitor. Note that like the scan lines produced in a TV monitor, not all of the manufacturer-quantified pixels used in the array produce the active image, and, similarly, several are reserved for signal transfer and picture synchronization timing, etc. Therefore, you will often encounter two sets of pixel numbers defined for a given array, being total number of pixels and total active pixels.

Considering that a progressive scan CCD reads out the entire image immediately, we can see in the case of a dual exposure interline transfer camera that should the position of a subject shift dramatically between each 1/50-second exposure, the resulting full-frame image may appear slightly smeared. On close inspection of a single frame of video, this may be seen as a jagged appearance around the borders of a star or a half-lit lunar crater where differences in contrast may be dramatic. But this can be corrected using a deinterlacing filter either during postimage processing or as a real-time filter, if your video capture software allows. Either way, its application will bring both fields back into correct registration, removing the jagged scan line appearance.

1.5 Video Resolution

Since pixels are micron-sized detectors, the smaller that manufacturers can make them, the more they are able to fit into an array for producing an image of higher resolution. As any camera shop employee will tell you, the greater the number of pixels a camera has, the finer the detail revealed in the resulting picture.

By today's standards, a typical CCTV camera, like that used for astronomy, is considered a low to medium resolution device when compared to the megapixels scale images of large format CCDs in direct digital output imagers such as a DSLR or SBIG astronomical cameras, for example. A digitized frame of full resolution interlaced video is limited to 768×576 pixels at best, using an economical capture device. But it is important not to compare apples with oranges because interline transfer video camera resolution is governed not just by pixel numbers in its CCD array but primarily by the number of picture scan lines it produces. Therefore, the greater the number of scan lines, the finer the detail or picture resolution.

For example, most basic surveillance-type cameras are quoted as producing roughly 320 TV lines, which for general use is quite good and better than the standard recording and playback resolution of a conventional 240 TV line VCR (video cassette recorder). But some offer higher resolutions of up to 600 lines or greater, being far better than 400 TV line Super-VHS recording quality. To view and obtain recordings of the maximum vertical resolution of the camera, a display and recording device that equals or exceeds the cameras specifications are required. This could mean the difference between detecting and recording a faint partner star during a lunar or asteroid occultation event or not (Fig. 1.6).

1.6 Low Light Performance

The light sensitivity of a CCD video camera is rated on luminance flux (lux scale) being the minimum amount of light or photovisual radiation (centered at around 550 nm) needed to produce a signal that can be interpreted to create an image at a given f/ratio. Early camcorders, for example, had lux ratings of around 10, but this has improved substantially over the years with many basic low-cost monochrome (black and white) cameras achieving 0.05–0.0001 lux or better. So a lower lux rating will reveal fainter stars, and this is one of the key features that has allowed the video camera to step up to the plate as a practical medium for deep-sky imaging. Furthermore, depending on the quantum efficiency of a particular design, CCDs with larger pixels collect more photons than those with smaller dimensions, thus yielding faint detail more efficiently (Fig. 1.7).

For all their benefits, CCD cameras are not devoid of irritants. The aesthetic appeal of the individual images they produce can be affected by a problem known as "dark current." This is a random, unwanted, noisy artifact caused by a build up of charges in each pixel due to heat generated within the CCD itself, supporting circuitry, and external temperatures. You can see this by closely examining a single captured frame of video. The image appears speckled or grainy throughout. This is why "still" image astronomy cameras are specifically designed for thermally cooled operation at around 20 or 30°C below ambient in order to greatly reduce this background noise.

Video camera manufacturers rate noise performance as a ratio measure of signal strength (the good stuff) relative to background noise (the bad stuff); this is often described as S/N or SNR (signal-to-noise ratio) and is measured in decibels. A camera rated ($S/N > 45$ dB) is desirable, and the higher the number, the better. The SNR ratio is closely associated to a CCD camera's practical dynamic range, and so the higher the signal response is over the unwanted noise, the greater its ability to record the maximum grayscale contrast steps. Some newer Mintron-based cameras and others now include excellent internal noise management filters known as SDNR (super digital noise reduction), which can tackle much of the problem at the pre-output stage.

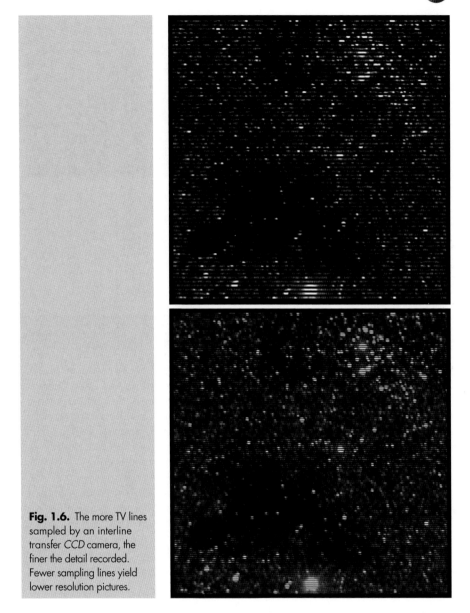

Fig. 1.6. The more TV lines sampled by an interline transfer CCD camera, the finer the detail recorded. Fewer sampling lines yield lower resolution pictures.

However, a thermally cooled video camera is not essential for producing aesthetically pleasing pictures. In fact, most amateurs today employ a post-processing technique called "image stacking" to reduce randomly generated noise, including other momentary induced artifacts such as RFI (radio frequency interference) while also accentuating the desired signal. We will cover this topic in more depth later.

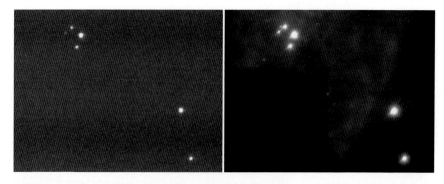

Fig. 1.7. Without any processing, this image of the Trapezium region in Messier 42 was taken with a standard sensitive monochrome *CCTV* camera (*left*) and the GSTAR-EX camera in non accumulation mode. Thanks to Sony's ExView HAD *CCD* sensitivity, note how the image on the *right* reveals nebulosity around the stars even in real time.

1.7 Hot Pixels

Other undesirable artifacts caused by heat include "hot pixels." These appear in an image or on a display monitor as pinpoint or star-like white spots and occur in almost all mass-manufactured CCD cameras. In astronomy they are sometimes referred to as false stars. Hot pixels are individual sensors with greater than normal operating rates of charge and leakage, most visible in long exposures. Most are faint, adding to the background noise and sometimes winking on and off randomly, while a handful may appear very bright or perhaps permanently visible.

As ambient temperatures rise, so, too, will the number of visible hot pixels, often increasing in brightness. In the case of general live viewing with a frame accumulation camera, thermally cooling it will yield a more aesthetically pleasing result on a video display. But, if you intend to create a deep-sky portrait without the misleading presence of false stars and using a computer to capture and process images then you will most certainly want to employ a postprocessing technique known as dark frame subtraction for removing hot pixels. This will also be discussed in more detail later.

1.8 Color Versus Black and White

It would seem logical and perhaps very appealing for the time-deprived astrophotographer to simply use a color camera to produce a few nice pictures of a galaxy or planet during that weekend away in the country. However, in the case of low/medium resolution cameras there is another side to the issue that is well understood by experienced amateurs, and that is spatial resolution.

All CCD image sensors are, at their core, monochrome detectors. The picture produced by the CCD in nearly all mass-produced cameras is an 8-bit image, made up of 256 shades of gray from pure white (255) to black (0). This is known as the dynamic range of the CCD, which is closely related to pixel well depth, a feature that determines the minimum amount of contrast detectable. In more expensive 10-bit cameras, the number of discrete grayscales can range up to 1,024.

Cameras that produce a color picture do so by selectively filtering out the primary visual wavelengths of white light that make up the colors we see in a rainbow, for example. These colors are RED (700–600 nm), GREEN (600–500 nm), and BLUE (500–400 nm).

Single CCD color cameras most commonly achieve this effect by placing rows, columns, or a matrix of microsized red, green, and blue filters across the image sensor which, when combined electronically, produce an image with the necessary shades and hues interpreted by our eyes and brain as color. Some CCDs may use a matrix of cyan, magenta, and yellow filters to achieve a similar outcome, but either way, only a third of the effective pixels in the array (those used directly for detecting photons) are used to produce each color, thereby decreasing overall spatial resolution. This is why quality camcorders utilize three different CCD sensors, thereby focusing specific color wavelengths to each via a dichroic prism in order to preserve maximum resolution. In terms of economics, this is why many amateurs prefer monochrome cameras and take individual exposures with specifically tuned red, green, and blue filters. Thus spatial resolution is maintained, and each filtered 8-bit monochrome image can later be combined to create a 24-bit color picture using image processing software.

Another point to note is that a purely monochrome camera has far greater light sensitivity than an equivalent color model, making them more desirable for imaging, particularly, and live viewing with small aperture instruments (Fig. 1.8).

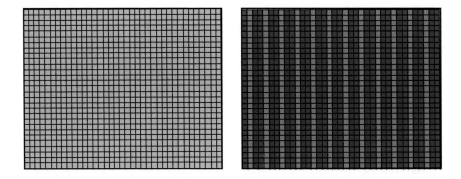

Fig. 1.8. Whether using a matrix or rows of color filters over the pixels in a CCD array, spatial resolution is sacrificed. Individual filtering of the entire monochrome array (*left*) is the preferred method for creating color images.

1.9 CCD Sizes and Image Scale

In terms of the physical CCD array sizes used in typical video cameras, they are most commonly defined using approximate diagonal measurements of 1/4-inch (4 mm), 1/3-inch (6 mm), and 1/2-inch (8 mm). At a given effective focal length the size of the chip also has a direct effect on the scale or apparent size of a subject being imaged (Figs. 1.9 and 1.10).

The size of the image sensor also affects how a particular lens will perform in terms of the field of view (FOV). Many CCTV-type lenses are made for use with standard 1/3- and 1/4-inch CCDs, so when used with the often preferred larger format 1/2-inch CCD, they can reveal unwanted vignetting (a circular fading of the image toward the edge of the FOV, like a tunnel effect). So if you want to use a CCTV lens, such as an auto-iris model with a 1/2-inch CCD camera for wide field imaging, then you need to check that it is specifically designed for this format.

When used in conjunction with a telescope of the same effective focal length, these different array sizes can be likened to interchanging eyepieces of different powers. In practice, the smaller the CCD, the narrower the field of view and the larger the planetary nebula or galaxy will appear on the monitor. Inversely, the larger the CCD array, the wider the field of view, presenting the object of interest somewhat smaller. Since planets have such a small angular size, cameras with 1/4-inch and 1/3-inch CCDs are preferred, while those with 1/2-inch arrays are best all around for deep-sky video imaging. Compared to larger format cameras this might still seem a little small.

But consider this: almost all the realistically accessible deep-sky objects we can observe and image make up more than 95% of the targets and have angular sizes that fit well within the array of a 1/2-inch CCD using the most common telescopes. Furthermore, detailed structure can be more readily observed and imaged with the larger-scale views produced by smaller arrays, as you will see illustrated throughout this book (Figs. 1.11 and 1.12).

A 1/2-inch CCD equates approximately to the magnification of a 7 mm Orthoscopic or Kellner eyepiece with a 45° apparent field of view. If you want to calculate the approximate field of view in arc minutes for your camera and telescope, you can derive it from the following basic calculations.

Image scale (seconds of arc/pixel) = Pixel size (μm)/Telescope focal length (mm) × 206.

FOV (minutes of arc) = Image scale × No. of pixel (on a given axis)/60.

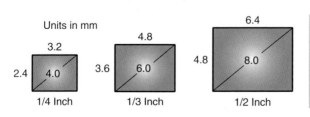

Fig. 1.9. *CCD* array sizes found in various video cameras.

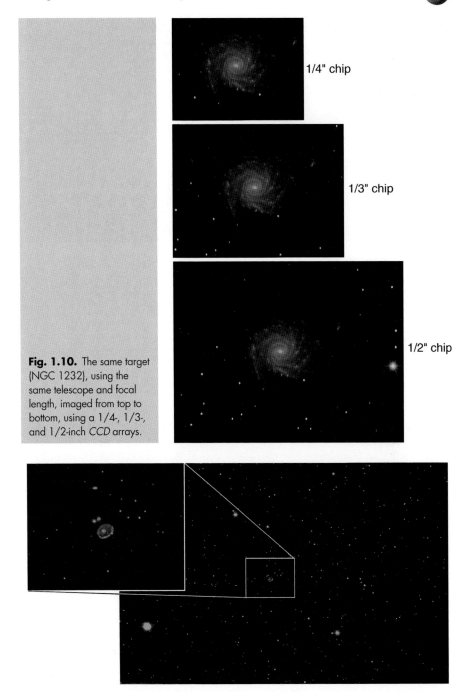

1/4" chip

1/3" chip

1/2" chip

Fig. 1.10. The same target (NGC 1232), using the same telescope and focal length, imaged from top to bottom, using a 1/4-, 1/3-, and 1/2-inch CCD arrays.

Fig. 1.11. The "Cartwheel" galaxy imaged with 35 mm color film and a GSTAR-EX Video camera with a 1/2-inch CCD chip, both through a 31.5 cm f/4.5 Newtonian reflector. The video CCD provides a greater image scale which reveals so much more detail.

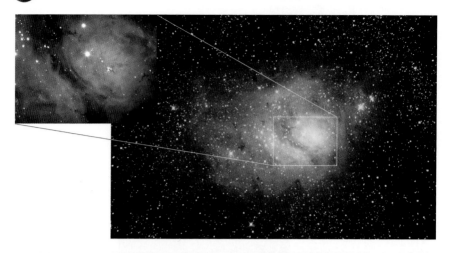

Fig. 1.12. Although a video CCD will not fit all of Messier 8 on the array, it does reveal beautiful detail in the area around the "Hour Glass" nebula compared to the 35-mm film using the same 31.5 cm f/4.5 Newtonian reflector.

Table 1.2. Field of view and magnitude limits (at full sense-up) for a 1/2-inch CCD of the GSTAR-EX camera.

Optics	Field of view	Approximate magnification	Real-time onscreen limiting magnitude
31.5 cm f/4.5 reflector	15' × 12'	200×	16.9
28 cm f/10 Schmidt-Cassegrain	8' × 6'	400×	16.2
25 cm f/5 reflector	18' × 13'	178×	16.2
20 cm f/10 Schmidt-Cassegrain	11' × 8'	286×	15.5
20 cm f/5 reflector	22' × 16'	140×	15.8
20 cm f/4 reflector	28' × 20.5'	114×	16.2
15 cm f/5 reflector	30' × 22	107×	15.6
8 cm f/5 refractor	55' × 41.5	57×	14.2
300 mm f/4 lens (7.5 cm aperture)	1° 15' × 55'	42×	13.6
135 mm f/2.8 lens (4.8 cm aperture)	2° 42' × 2° 03'	20×	12.8
8 mm f/0.8 lens	43° × 33°	1.1×	8.5

Table 1.2 will give you an idea as to the field of view achievable with various telescopes and lenses we have tested. The magnitude limits are from a country sky and may be slightly less from city skies.

Although aperture is the key to gathering more light delivered to the CCD, shorter focal ratios also make a big difference in helping to detect faint stars and extended objects. Long focal lengths are best suited to small objects with high surface brightness, such as bright planetary nebulae. For most extended objects, focal ratios from f/8 to f/4 are best. Ratios shorter than f/4 will place higher demand on the optics to

Fig. 1.13. The famous Andromeda Galaxy (Messier 31), imaged with a single-lens reflex (*SLR*) lens of 135 mm focal length @ f/2.8 onto a 1/2-inch chip video camera. Field of view equals 2.7 × 2°.

FOV 43 x 33 degrees   Comet McNaught 2006 P1   Steve Quirk 2007 01 22   GSTAR-EX 8mm lens LRGB

Fig. 1.14. Comet McNaught (C/2006 P1), imaged with a C-mount security lens of 8 mm @ f/0.8 onto a 1/2-inch chip video camera. Field of view equals 43 × 33°.

be in perfect collimation and may show some distortion due to coma around the edge of the field of view.

Use of a coma corrector, like the model designed specifically for Vixen's fast R200SS f/4 Newtonian or generic models such as the Baader MPCC, used with many different telescopes, may be required. Figures 1.13 and 1.14 show a couple of examples of wide field targets that are well suited to short focal length lenses.

CHAPTER TWO

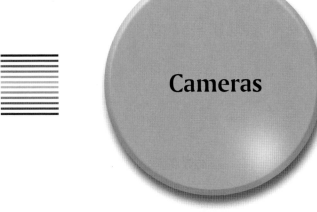

Cameras

We all have some expectations when making the decision to purchase a new gadget that is talked up by a vendor or by another amateur who may possess completely different equipment, skills, and experience. There is nothing worse than realizing later that what you bought did not meet those expectations. So, in this chapter, we will look at cameras capable of producing deep-sky images and the various essential functions and features that are important to ensuring they are up to this task.

2.1 Image Accumulation Cameras

The great innovation in video cameras for use in deep-sky astronomy has been the development of short-exposure image accumulation technology. This internal exposure co-adding technique makes it possible to observe faint detail otherwise nearly impossible to see at the eyepiece and to make a deep-sky portrait from light-polluted urban areas. Moreover, the short exposures mean less stringent need for perfectly polar-aligned mounts or correcting periodic tracking errors. Providing the amount of image drift is small and occasional recentering of the target is carried out, then images can later be aligned and stacked using popular freeware programs such as Registax.

Among the handful of CCTV manufacturers utilizing this technology, perhaps the two most prominent innovators are Mintron in Taiwan and Watec in Japan. Both have developed such cameras to meet the growing needs of the security and surveillance industry for operation in extreme low-light situations. By using the latest in highly sensitive image sensors such as Sony's ExView HAD, combined with

S. Massey and S. Quirk, *Deep-Sky Video Astronomy*,
DOI: 10.1007/978-0-387-87612-2_2, © Springer Science + Business Media, LLC 2009

Fig. 2.1. Set to accumulate frames, the Moon's Earthshine shows as though fully lit by the sun. Several faint stars can be seen in the background sky.

smart onboard image processing, these new generation cameras produce highly intensified images (Fig. 2.1).

In both manufacturers' models, the number of exposures accumulated within the cameras' internal memory buffer before output is manually selectable using incremental steps from X2 upward. Watec has a model it specifies as its astronomical-imaging camera. The WAT-120N allows frame accumulation rates of X256 and greater, but unlike Mintron cameras, the picture output remains blanked until full accumulation status is reached.

In Mintron cameras the incremental frame accumulation function is known as "SENSE UP," which is also quoted in their sensitivity specifications as StarLight Mode. In other manufacturer models the accumulation function or control is simply referred to as accumulation.

Each individually accumulated image is updated, or rather refreshed, at the camera's output within a specific amount of time according the number of frames being co-added as selected by the user. Once maximum accumulation is reached (taking up to a minute or so), you can simply move to any deep-sky object and see it instantly on the monitor without having to reset this function. When doing so, you will notice a streaked appearance of stars-like meteor trails across the viewing monitor until telescope slewing has been completed. The length of the trails is directly related to the accumulation / exposure time and telescope slew rate.

Another recent entry into the frame accumulation camera market is a company called The Imaging Source. They produce a very good range of compact, progressive scan, Firewire IEEE 1394 and USB 2.0-interface cameras which, combined with their own software, are capable of short and long exposures. They have even tailored certain models specifically for use in astronomy, with frame rates from 1/60 of a second and integration times up to 1 hour! But of course, longer integration time means slower refresh rates, limiting your ability to monitor the progress of each image being recorded. So, you will certainly need a quality tracking mount or autoguiding system to avoid drifted star images when using these extended exposures.

Cameras

The Imaging Source cameras utilize some of Sony's best color and monochrome square pixel, progressive scan image sensors in 1/4-, 1/3-, and 1/2-inch formats. Internal filters effectively manage unwanted thermal background noise very efficiently, particularly for an uncooled camera. The Imaging Source color cameras are particularly good in applications where the user may not want to fuss with tricolor filtering and wants to produce an attractive color image in one simple step (Figs. 2.2 and 2.3).

Fig. 2.2. The lightweight DBK21A-FO4.AS Firewire color camera by the Imaging Source is capable of fast exposures and on-board image accumulation.

Fig. 2.3. Messier 42 imaged with the DBK21AF04.AS through a 100 mm f/7.7 ED refractor and Vixen 0.6X focal reducer. Twenty-four stacked images from 5 second accumulation exposures.

2.2 Cameras Modified for Astronomy

Although companies like Santa Barbara Instrument Group (SBIG) (in the United States) produced innovative video systems like their STV unit specifically for use in astronomy some years ago, a number of astronomical equipment suppliers have taken steps to tailor the aforementioned CCTV manufacturers' cameras for use at the telescope. In particular, companies such as Adirondack in the United States, SAC Imaging in Europe, and Binary Systems in Australia have developed hand controllers, cooling systems, and astronomy-specific software for cameras, such as the StellaCam range and the GSTAR-EX. These cameras come standard with C/CS-threaded rings to take specific C-mount adapters and CS-mount lenses commonly used for meteor work and general night-sky surveillance. For use with a telescope they are supplied with various C-mount adaptors such as 1.25-inch nosepieces that are threaded to accept standard accessories such as colored glass Wrattens, Red, Green and Blue (RGB), and infrared (IR)-blocking filters. Other optional items like C-mount focal reducers and tele-extenders are also available.

Indeed, some suppliers custom modify camera electronics to allow for an even greater number of accumulated exposures, but this can lead to more unwanted noise and hot pixels, so cooling the camera becomes a mandatory, cost-adding requirement. Furthermore, output refresh rates are also longer (i.e., 10 seconds and up), so one might ask why not just simply use a standard cooled CCD imager like an SBIG or Starlite Express? A good argument, but on the contrary, there remains the ability to record directly to a VCR and view directly on a standard TV monitor. Furthermore, some cameras allow "only" accumulated picture output, lacking the versatility of "standard" video shutter speed-controlled imaging useful for doing lunar, planetary, and occultation work when required.

One of the goals of this book is to show what can be achieved with the most basic type of frame accumulation video camera without all the bells and whistles, so you can make an educated decision on what will best meet your needs. With this in mind, for much of this book we have used a GSTAR-EX camera to give you an idea of how basic accumulating video cameras (without costly modifications) can perform as a wonderful, simple, and affordable tool for deep-sky imaging (Figs. 2.4–2.6, Table 2.1).

2.3 Basic Camera Features

We have talked about the scanning systems for PAL and NTSC, pixels and video resolution, and lux sensitivity. Because having a well-rounded understanding of the camera's controls will ensure you can obtain the best raw image output the camera can deliver for a given situation, we will now briefly look at the most important camera functions for achieving the best results.

Fig. 2.4. The GSTAR-EX camera with optional hand control.

Fig. 2.5. StellaCam 3 fitted to a Newtonian telescope. Courtesy of Phillip Hinkler.

Fig. 2.6. Watec 120N-Plus image of the keyhole at the heart of the Carina Nebula using a 10-second accumulation time. Courtesy of Allan Gould.

Table 2.1. Some typical specifications and features you may encounter.

Important	
TV system	CCIR (PAL) or EIA (NTSC)
Image sensor size	May be 1/4-inch, 1/3-inch, or 1/2-inch depending on camera
Total CCD pixel no.	795(H) × 596(V) CCIR – 811(H) × 508(V) EIA
Scanning system	625 lines CCIR – 525 lines (EIA)
Sensitivity (Lux)	0.02 normal operation/0.0002 accumulation mode @ f/1.4
AGC (auto gain control)	Automatic and manual functions
Electronic shutter speed	1/50 (CCIR) – 1/60 (EIA) s to 1/12,000 s
Frame accumulation mode	2–128×, 256× or 512×
Gamma correction	0.45/1.0 selectable
Signal to noise ratio	52 dB (minimum) with AGC set to OFF
Video output	Composite (BNC) 1.0 V peak to peak 75 Ω and/or S-Video
Power requirement	Typically +12 VDC (center positive)
OSD menu	On-screen display menu system
Nice extras	
Grayscale calibration bar	ON and OFF selectable for calibrating your monitor
Digital zoom	2× or more to assist with initial telescope focusing
Mirror function	Nice for flipping images to correct sky view in certain telescopes

CCIR Comittee Consultatif International Radiotelecommunique, *PAL* phase alternating line, *EIA* Electronics Industry Association, *NTSC* National Television Systems Committee, *CCD* charge coupled device, *BNC* Bayonet Neill-Concelman

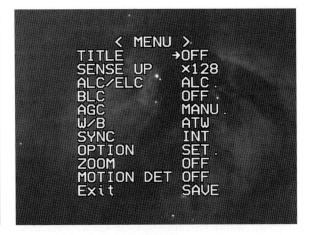

```
        < MENU >
  TITLE        →OFF
  SENSE UP      ×128
  ALC/ELC       ALC .
  BLC           OFF .
  AGC           MANU .
  W/B           ATW
  SYNC          INT
  OPTION        SET .
  ZOOM          OFF
  MOTION DET    OFF
  Exit          SAVE
```

Fig. 2.7. Handy on-screen displays are useful for making changes to camera settings while still viewing the deep-sky subject in the background (NGC 2024).

2.4 Camera Control

Camera functions can vary, depending on the manufacturer's design and dealer product customization. For example, settings may be accessed via physical buttons on the camera itself or a hand control. The GSTAR-EX has an array of push button switches on the back panel, but like a StellCam III, it can be controlled via an optional hand controller. Some cameras can even be controlled via an auxiliary port using an optional RS232- or RS485-interface cable connected to a computer. For making changes, these cameras have an on-screen display (OSD) menu system that still allows you to see what is going on with your targeted object in the background while you enter submenus and make adjustments. Some cameras simply provide rotary switches on the hand control for adjustments to exposure, signal gain, and so on (Fig. 2.7).

One very handy feature of the GSTAR-EX is its mini-din RS232 auxilliary connection which also doubles as the optional hand controller interface. It allows for remote control of the cameras primary functions using free camera interface software and a special cable connected to a computer.

2.5 Shutter Speed

The term shutter speed heralds from the old mechanical system in film cameras. Like a curtain opened for a specific time, then closed to stop any further light from reaching the film, the duration of open time combined with the aperture setting directly affects how much the film is saturated by photons.

The exposure time in a CCD camera is managed on the same principle, although by electronically switching the array on and off. The maximum exposure time is 1/50 of a second for CCIR (PAL) cameras and 1/60 of a second for EIA (NTSC). The duration times for the "on" state are then selectable as shorter or faster exposure

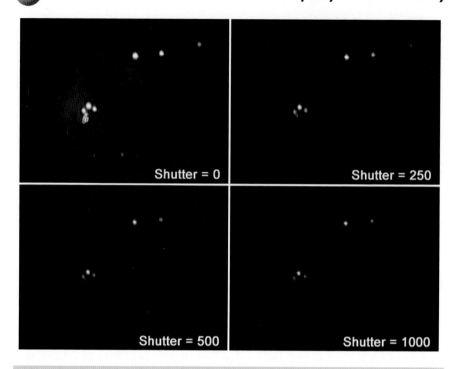

Fig. 2.8. The trapezium at the heart of Messier 42 used here to show the effects of various shutter speeds when frame accumulation mode is set to OFF.

times: 1/100, 1/250, 1/500 of a second, and so on. The shorter the exposure time the lesser the time available for light to build up in the array, thus producing progressively darker images with each step of faster shutter speed used. These standard camera shutter speeds are useful for solar, planetary, and lunar work in particular, as well as occultation imaging and timing (Fig. 2.8).

2.6 Signal Gain Control

Often called AGC (automatic gain control), it is preferable that the user be able to manually adjust this function as well. This signal control plays a significant role in revealing the faint stuff we seek as astronomers in extended objects, such as nebulae and galaxies. AGC is, in essence, a signal-boosting amplifier, but in doing so it also increases unwanted background noise and the intensity of hot pixels. In other words, it amplifies the entire signal detected and output by the CCD.

In deep-sky imaging the use of AGC at maximum is generally mandatory, and the unwanted noise it amplifies can later be minimized by stacking images. In conjunction with the exposure accumulation rate setting, manual adjustment of signal gain can be the defining factor between a burned out core in the nebulosity around the Trapezium stars of the Orion nebula or defining them as individual stellar objects. It can mean the

Fig 2.9. Left image of NGC 3324 in Carina is full camera gain and normal software gain. Right has the software gain increased to easily show much more nebula in the live view.

difference between detecting a faint comet or not or seeing a very faint nebula or other object live on the monitor beyond that which can be seen through the eyepiece.

By increasing the gain further with your capture software, very faint objects can be seen on screen. Although the view will be much noisier, the object will be revealed positively. This mode of operation can really help when trying to find or identify a very faint target. Then change back to a standard or less noisy setting for the image-recording session (Fig. 2.9).

2.7 Gamma Settings

The faint detail we all seek in features like the wispy festoons in the cloud tops of Jupiter or the spiral arms of a galaxy mostly reside in the middle gray regions of an image. This detail is often difficult to reveal using only the brightness and contrast adjustments of image-capture software drivers. Gamma adjustment can be regarded as a supplementary contrast enhancement that can yield improved visibility between darker gray levels (where outer spiral arm information exists) and brighter levels like stars or the central core of a galaxy, thus enhancing the overall pixel dynamic range. Some cameras have built-in 10-bit A/D analog to digital converters that provide better contrast or gamma correction in the image by sampling two extra bits of data, effectively yielding four times more discrete shades of gray.

The gamma control is a switchable or variable function usually ranging from 1.0 (no difference) to 0.45 (widely stretched mid tones), the latter being the desired setting. A manually adjustable stepped setting is very desirable, providing a more flexible control over the amount of total mid-tone brightness level adjustment.

Gamma adjustment is also one of the most important tools in post-processing of images to reveal the faintest middle gray tone detail in an image. This will be explained further in our post image-processing chapter (Figs. 2.10 and 2.11).

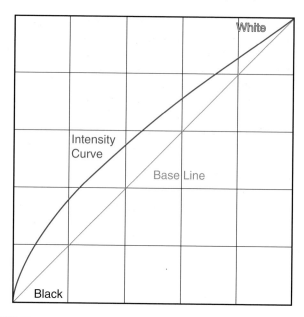

Fig. 2.10. Gamma curve chart showing how the mid-brightness regions are increased in value while the black and white ends are left normal.

Fig. 2.11. These images of Comet P/17 Holmes were taken with gamma setting of 1.0 (*left*) and 0.45 (*right*). 135-mm lens and 200 video frames with GSTAR-EX.

2.8 Accumulation Mode

Now we come to the most important function that makes the inherently short exposures of a video camera capable of recording faint deep-sky objects, the frame accumulation mode. Not to be confused with long exposures achieved with extended open shutter/array on times, this is a simulated long exposure and is based on the number of co-added exposures taken at the maximum video exposure capability time of 1/50 of a second for CCIR (PAL) or 1/60 of a second for EIA (NTSC). In other words, the exposure time itself is no greater than 1/50 of a second; however, the camera electronically accumulates and co-adds a user-defined number of these short exposures to yield the equivalent result of a picture produced by a "still" picture-integrating camera capable of taking "true" longer-duration exposures. The user selected setting for the number of exposures being co-added within the cameras memory buffer determines how fast the picture is refreshed at the camera's output as seen on a video monitor.

How exposures are treated in the accumulation process varies from one camera to another, depending on its circuitry and programming. One process is to co-add each interlaced field and use interpolation techniques to generate the second field, making up the entire video frame or image. Other models may co-add both unique fields, resulting in longer picture refresh times but with improved resolution (Fig. 2.12; Table 2.2).

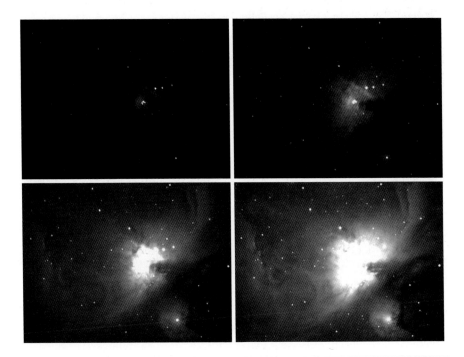

Fig. 2.12. Center of Messier 42 shown with no accumulation used then X8, X48, and X128 sense up.

Table 2.2. A guide to accumulation rate versus screen refresh times (PAL) for a basic video accumulation camera using field interpolation.

Accumulation mode	Picture output refresh rate (in seconds)
X2	0.04
X4	0.08
X6	0.12
X8	0.16
X12	0.24
X16	0.32
X24	0.48
X32	0.64
X48	0.96
X64	1.28
X96	1.92
X128	2.56
X256	5.12

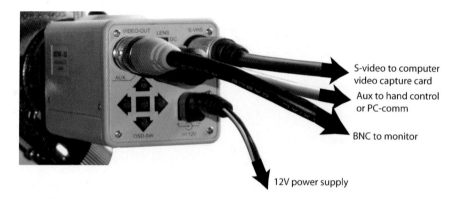

S-video to computer video capture card

Aux to hand control or PC-comm

BNC to monitor

12V power supply

Fig. 2.13. Back panel connections of the GSTAR-EX camera.

2.9 Video Cables and Connections

While CCD surveillance-based cameras most typically used in astronomy process internal signals digitally, the standard analogue signal amplitude from camera to interconnecting devices is 1 V peak-to-peak into 75 Ω (Ohms). The video signal may be either Composite [using Radio Corporation of America (RCA) yellow socket or chrome-plated twist and lock Bayonet Neill Concelman (BNC) style connector] or Y/C separated S-Video [a 4-pin Deutsches Institut für Normung (DIN) connector].

Disregarding all the timing, bandwidth, and other aspects encoded into a video signal, which are mainly of interest to TV technicians, composite video is essentially a modulated signal containing the (Y) luminance and (C) chrominance information. This signal must then be demodulated or decoded by a receiving device such as a VCR or a video monitor. But it is considered that losses to signal integrity can occur during this encoding and decoding process, particularly over long distances. S-video connections, on the contrary, pass the Y and C components along separate wires in order to maintain source (camera) signal quality that might otherwise be compromised as a combined modulated signal.

For monochrome cameras, where only a luminance signal is produced, a visually detectable difference in quality between composite and S-video is pretty much indiscernible in the "real" world when it comes to creating your aesthetically pleasing deep-sky portraits.

It is best to maintain the shortest cable length possible, since long cables can act like an antenna and become more prone to external signal interference over increased distance. And since copper wire is not a perfect conductor, the voltages required to maintain good signal integrity are reduced by resistance over longer distances. The thickness of the conductors and quality of the braided shielding inside the cable will govern the maximum practical length before the signal starts to noticeably degrade. Lengths of up to 15 m with quality 50 Ω cable (typically used for general consumer audio and video connection) is considered acceptable. Much longer distances can be achieved with well-shielded 75 Ω coaxial video cables. It is also vitally important to keep video cables well separated from other cables, such as shared AC mains power, where switching in the motors of a fridge, washing machine, or clothes dryer, for example, can create unwanted noise spikes. Poorly shielded equipment with microprocessors or electronic switching devices like the motor drives of a telescope mount can also induce unwanted noise.

When it comes to USB and Firewire cameras, the main advantage over analog designs is that power is provided from the computer, meaning one less cable and plug pack power supply you need to juggle with at the telescope. On the down side, high-speed digital interfaces, such as these, limit the workable distances between camera and telescope. Depending on the camera, most USB and Firewire leads are limited in operation from 2 to 3 m. However, in our trials using one Imaging Source Firewire camera, a cable of 5 m worked without error. If you need to achieve greater digital cable lengths for more remote use, there are affordable "active" extension cable systems available to maintain digital signal integrity over longer distances. You should check the specifications to ensure they have nickel plated corrosion-proof connectors and gold-plated contacts for maximum conductivity with twisted pair impedance matched wiring to avoid electromagnetic or radio frequency interference (Fig. 2.14).

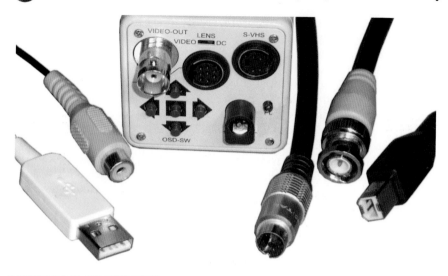

Fig. 2.14. Various camera connector interfaces including composite, S-video, and digital.

CHAPTER THREE

Video Capture

For many videographers, the primary goal is to obtain a deep-sky portrait using a computer that can later be posted to a Website, printed to glossy paper, or submitted to a magazine.

3.1 Digitizing Video

If your camera already has a direct digital output, such as a USB or FireWire interface, then use of a third-party computer capture interface is not required. But for analog output cameras, decoding the composite or S-video signal to a computer requires an analog to digital converter (ADC), commonly known as a video-capture device. Very affordable these days, ADCs range from video-to-USB converters suitable for most computers or dedicated peripheral component interconnect (PCI)-interface cards that can be inserted in any spare PCI slot on the mother board of your computer. There are also PCMCIA plug-in devices for laptops that are often preferred for their sturdier connection and onboard processing power over simple inline USB converters. Most of the low-cost models are designed specifically for people wanting to create home movies, so are usually provided with a suite of software for enabling the user to capture footage and import it into a movie-editing program, then add effects, and finally burn the finished movie to a blank DVD.

No matter which device you intend using, you should first ensure that it can accurately reproduce images from your camera and at the maximum resolution possible. In most economical capture devices these days, the digitizing chipset and associated drivers supplied with it determine the maximum picture scale (image pixel window size)

S. Massey and S. Quirk, *Deep-Sky Video Astronomy*,
DOI: 10.1007/978-0-387-87612-2_3, © Springer Science + Business Media, LLC 2009

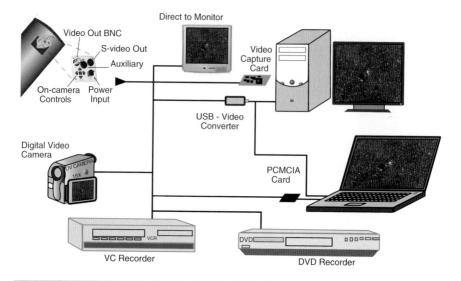

Fig. 3.1. A variety of recording and monitoring devices that can be connected to a video camera.

you can capture. Common full-resolution video-capture window formats generally fall in line with requirements for super video compact disc (SVCD), DVD, and DV formats, such as PAL 720(H) × 576(V) pixels or NTSC 720(H) × 480(V). Some low-cost USB video-capture devices cannot cope with or feature full resolution frame capture at rates higher than 15 frames per second (fps), and although this does not present a problem for deep-sky imaging, it does limit your use of the device should you want to capture lunar and planetary targets using maximum desirable rates of 25–30 fps. Ultimately, it is better to spend a little more now for something that will cater best to both needs.

How the scan lines of a video composite signal are converted to an array of pixels in an image of a given aspect ratio (either 4:3 or 16:9) is a highly convoluted topic and a process beyond the scope of this book. But in short, much of it concerns the handling and conversion of non-square pixels [like those used in CCDs of typical CCTV cameras] to square pixels produced on a computer display. In essence, we simply want to get a correctly proportioned image at the highest possible resolution on to a computer (Fig. 3.1).

3.2 Software Capture Tools

What product you choose to capture the video with from your camera is always a personal choice based on a balance of quality, features, and cost. Your capture device will almost certainly come with a disk of packaged programs, including the all-important capture software and drivers that provide a live view window on your computer, but

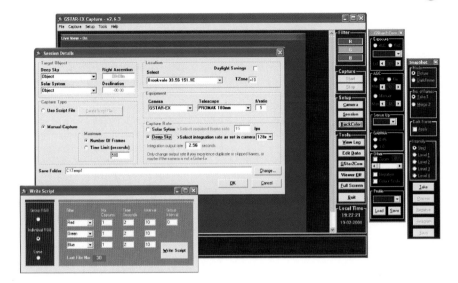

Fig. 3.2. Screen shot of the popular GSTAR capture software showing various submenu features.

most are limited to a slowest capture rate of 1 fps. Among the huge range of capture devices available, those such as Leadtek's *WinFast* T.V2000 XP Expert or *Smart Capture* by Innodv are very good PCI-based products with bonus suites of supporting software for general use but still with a limiting slowest capture rate of 1 fps. There are also many freeware programs available on the Internet, which are found by doing a simple online search using the words "video capture program." But since frame accumulation cameras can also output images at rates slower than 1 fps, software able to capture each unique frame of video without replication is uncommon.

Below, as a guide, is a table describing just a few of the programs out there and their features. There are many applications available, but these are a few we have found that work well. Do an Internet search using the product names to find the most current Websites from which you can download (Fig. 3.2; Table 3.1).

3.3 Aspect Ratio

Since most CCTV-based cameras, like the GSTAR-EX or StellaCam, have image sensors that produce pictures with a 4:3 aspect ratio, it is desirable to capture images to a computer within the same format. In other words, 16:9 letterbox-style images common to wide screen liquid crystal display (LCD) displays will give us a stretched and unnatural result on the more common 4:3 aspect display. Since, most conventional CRT monitors have a 4:3 aspect ratio, the desirable capture window sizes are directly related. For example, 640×480 [$640(H)/4 \times 3 = 480(V)$] and these evenly

Table 3.1. A few examples of available capture software tools and their primary features being suitable for use with Deep Sky capable video cameras.

GSTAR Capture (freeware)	This program has numerous features but is specially designed for capturing video at rates slower than 1 fps. It can capture a unique frame of video every 2.56 s or greater and features a two-frame image intensity enhancement snapshot for identifying comets along with LRGB-filtered image timers, auto deinterlacing, object and equipment database, crosshair reticles, user set capture time limits, and RS-232 remote camera-control interface, among many other features. It also includes post-AVI image selection and output from AVI to sequential bitmap images for stacking in other programs.
VirtualDub (freeware)	A longtime ongoing project for general video enthusiasts, VirtualDub has a capture routine compatible with most capture devices and can also digitize incoming video pictures at rates slower than 1 fps. It also includes post-capture routines for brightening, darkening, sharpening, video deinterlacing, and rescaling, among other things. A new AVI can be compiled and written to disk or individual images exported.
WinFast PVR (commercial)	Ignoring its PC TV features dependent on the installed capture card, the WinFast PVR is an easy-to-use application provided with WinFast multimedia video capture devices. The PVR capture application allows you not only to view and record live TV or video input sources but also has a neat progressive filter mode feature that effectively deinterlaces incoming video frames on the fly, yielding a smoother looking live and recorded image. This is especially noted in the case of large, bright stars. It is limited to 1 fps slowest capture rate, but its software drivers combined with the card chipset also allows customizable window capture sizes to full PAL video resolutions, such as 768 × 576 pixels.
WinAVI (commercial with trial version available)	Can capture from AV devices, such as webcams and VCRs, and save to your computer or burn directly to disc. It records in formats such as AVI, WMV, DivX, RM, DVD, VCD, SVCD, or MPEG1/2/4. There are, of course, numerous other programs available that provide similar features.

AVI audio-video interleaved, *PAL* phase alternating line, *DVD* digital video disc, *VCD* video compact disc, *SVCD* super video compact disc, *WMV* windows media video, *MPEG* Moving Picture Experts Group, *RM* RealMedia, *VCRs* videocassette recorders, *DivX* digital video express

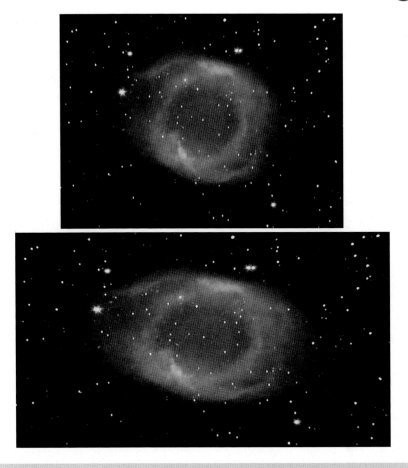

Fig. 3.3. Here we see a standard 4:3 aspect image compared to the abnormally stretched appearance of a 16:9 aspect ratio capture (NGC 7293).

rounded numbers further subdivide into other smaller 4:3 aspect picture sizes, such as 320 × 240 and so on. Using the above calculation you can see how the classic full resolution PAL (CCIR) picture scale of 768 × 576 fits neatly into the 4:3 aspect ratio requirement (Fig. 3.3).

3.4 Capture Resolution

It is important where possible to capture the maximum resolution we can from the video image produced by our deep-sky frame accumulation camera, particularly for printing purposes. Since we know from our discussion in Chap. 1 how traced scan lines are managed on a CRT display, we can also use this knowledge to better understand the image digitizing process from the intensity information contained within each video scan line output by the camera.

Converting the vertical rows of "active" scan lines in a video signal is a relatively straight forward process of digitally sampling each row produced from the top to the bottom into a matrix of pixels. In the case of a PAL (CCIR) camera, we, therefore, sample our 576 active lines. This includes the two half lines considered to make up one complete scan line of active image data. So this gives us our vertical aspect, but what about the horizontal? This is determined by the sampling clock frequency used by the capture device. The common sampling rate found in almost all video gear today as defined by the industry standard ITU-R BT.601 is 13.5 MHz; this is used by manufacturers in nearly all low-cost, mass-manufactured video-capture devices, even in DV camcorders.

Sticking to our PAL signal example and disregarding the timing signals contained in a single scan line, we are left with about 52 µs of the active scan line to sample. To convert this into pixel numbers given as a horizontal width, we can say 13.5 MHz × 52 ms = 702. But most capture devices sample at rates to yield a standard 704- or 720-wide format. We can see here that if we sample at a faster rate the signal could be interpreted into a few more horizontal pixels, such as the square pixel standard of 14.75 MHz. This translates as 14.75 MHz × 52 ms = 767 rounded to 768. But how a capture device achieves an image of this size may vary in terms of using true sampling frequency or by interpolating pixels. In reality, for general viewing purposes and aesthetically pleasing images, the effect of the latter in processing images is quite undetectable, so do not be too concerned.

So use the maximum capture frame size provided by your capture device where possible, being either PAL (CCIR) 768 × 576 (preferable) or 720 × 576 and for NTSC EIA 720 × 480. Images that do not strictly conform to the 4:3 aspect can later be rescaled on a desired axis, with image processing software to yield a more accurate on screen appearance (Fig. 3.4).

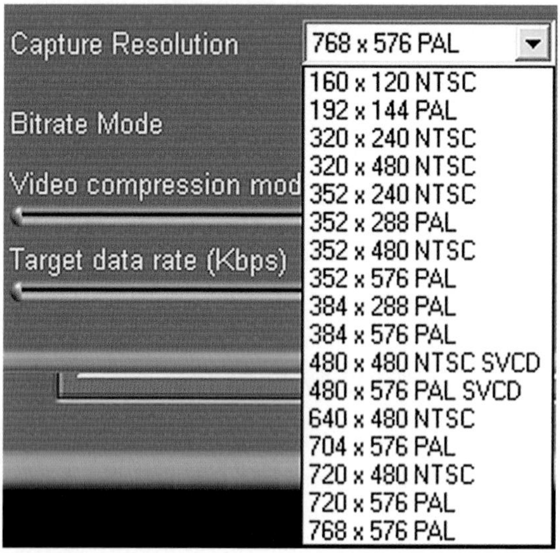

Fig. 3.4. Capture window sizes and matching TV standards provided by a commercial video-capture device.

3.5 Image Quality

When capturing a movie file to a computer, the most common format still widely used is the AVI or audio-video interleaved standard, and nearly all compatible capture devices support this long-established format, along with widely used video-processing programs like VirtualDub and Registax image stacking software.

Most capture devices offer a range of image codecs (code and decode) compression algorithms, which are designed to intelligently minimize the movie file size with the least possible effect on picture quality. Although much has improved in pixel manipulation compression techniques over the years, they are best used for personal webcam or family videos you want to post on the Internet, as there is unfortunately always a trade off in quality and raw pixel integrity vital to imaging in astronomy.

It is recommended that you should avoid use of video compression and stick with uncompressed AVI capture, if possible. But if you are using a color camera, then codecs like YUV are often preferred because they place less demand on computer processing for the color component of the signal while also not affecting the important luminance part of the image. (Y) relates to the luminance aspect of the picture and (UV) relates to the chrominance (color) part of the signal.

The image-compression format tells the video-capture device what "color space" to capture in. The options of image format range from 32-, 24-, and 15-bit RGB to YUV4:4:4, YUY2 (YUV4:2:2), YVU12, and YVU9.

The following is a simple guide to the effectiveness of different YUV codecs versus RGB if using a color camera.

3.6 RGB Color Spaces

RGB32: RGB32 provides the user the same quality as RGB24 with 8 bits for the red, green, blue, and alpha channel. It effectively yields 33% more information due to an unused alpha channel.

RGB24: This is said to have twice the horizontal color resolution as YUY2. This extra color resolution is usually produced via upsampling of the video signal. Most capture devices will not yield any additional quality during capture, and this format produces 50% more unnecessary data for our purposes. However, it is by far the most compatible format when it comes to video codecs for DVDs and playback on home theater systems.

RGB16: This provides the same data rate as YUY2 (33% less than RGB24), but there is significant banding in the signal. RGB16 is known to be good for video displays but poor for video archival.

YUV4:4:4: Within the YUV4:4:4 color-space formats, each of the three individual Y, Cb, and Cr components have the same sample rates. There is no compression of the signal components. This mode is sometimes used in cinematic post-production.

YUY2: (YUV4:2:2) The two chroma components are sampled at half the sample rate of luminance. This sampling is achieved by cutting in half the horizontal chroma resolution. There is a reduction in bandwidth of a video signal transmitted by approximately one third with little to no visual difference to YUV4:4:4. In most cases, this is the best algorithm, as its color space format is the closest to video data directly from the video decoder, and it offers significant speed/size improvements over RGB24 (it is 33% smaller).

UYVY: A byte ordering change from YUY2.

YUV4:1:1: In the color-space format YUV4:1:1, the horizontal color resolution is quartered. The bandwidth is halved compared to YUV4:4:4. This format is not considered of broadcast quality but is basically acceptable for low-end and consumer applications.

YVU12: (YV12/I420) YVU12 has 25% less data than YUY2. The trade off is in exchange for slightly more color bleeding vertically. There is no difference in luminance to YUV4:4:4.

YVU9: This holds 25% less data than YVU12 and 43% less data than YUY2 and significant color bleeding due to the chroma subsampling. YVU9 is not very well supported and is not recommended.

Considering that only the luminance and chrominance channels are utilized, the most desirable color space to operate in for video capture is the YUV format, specifically **YUY2**.

3.7 Recording Directly to DVD

If you are using a DVD recorder, then image compression is unavoidable, but we recommend that you select the "short play" best recording quality mode available from the DVD recorder set-up menu options. The same rule also applies to VCRs where short play/highest quality recording mode can be set by the user (Fig. 3.5).

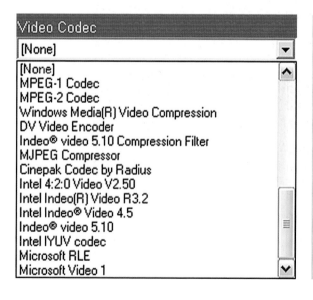

Fig. 3.5. Although we recommend avoiding use of image compression code and decodes (*codecs*), the typical selections from a capture device are shown here.

3.8 Bright Object Smearing in Videotape Recordings

Depending on the quality of your VCR, a common problem very prevalent in videotape playback is horizontal smearing from bright stars. This is the result of the large amplitude increase in the luminance component of a video signal that has not passed through a noise-reduction circuit or filter before reaching the video record heads. It appears like a translucent streak or smudged light extending to the right side of an extremely bright source. Most notably, a bright star or planet will reveal this when greatly contrasted against a dark sky background. The effect is similar to oversaturated pixel bleeding that occurs in some "still" image CCD cameras. So this can be a downside to using a VCR, unless it has a good noise-reduction filter built in.

Of course, the smearing artifact can be removed in postprocessing as long as great care is taken not to alter any true background detail. At the end of the day, it is simply better to avoid the problem from the outset if you can see it occurring in the playback tapes on your VCR (Fig. 3.6).

Fig. 3.6. Recorded on a budget VCR, two bright foreground stars near the galaxy NGC 253 show amplitude smearing in this stack of 25 frames from videotape. The smearing is seen as meteor-like trails extending to the right.

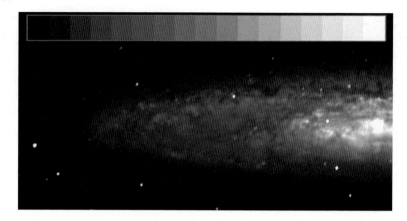

Fig. 3.7. The importance of stretching the middle gray regions of an image cannot be emphasized enough. Here we show a grayscale reference bar and how the various steps of gray contrast relate to revealing the faint detail in the galaxy NGC 253.

3.9 8-Bit Grayscale Images

After a capture device has sampled and quantized the analog voltage signals, the resulting 8-bit monochrome image is then made up of rows and columns (a matrix) of pixels. Every pixel in the image is assigned a brightness value from 0 to 255 and unique location address. More expensive monochrome cameras and capture devices with 10-bit and 12-bit processing yield a far greater number of discrete values, but for our purposes we will deal with the more commonly produced 8-bit image.

For faint fuzzies, such as galaxies and nebulae, nearly all the structural and interesting information lives in the middle gray areas. When increasing the contrast (bright and dark levels) of an 8-bit image we are effectively forcing the original intensity values into new values. So, lighter middle grays are pushed further to the white end of the scale, while darker shades are pushed to the black end. The image then becomes one with a narrower range of discrete gray shades to work with. In other words what started out with 256 discrete gray shades may now only consist of 128 or less. Furthermore, the less discrete gray shades there are in the image, the less photorealistic the picture becomes (Fig. 3.8).

This is important to understand not only for post-processing of your images but also at the time of capture because you will want to record the maximum range of gray values your camera can output in the raw images to work with later. So be sure to set your capture software driver brightness and contrast input levels to around the midway point before capturing your video. Another important point is to be mindful of the fact that the night sky is not completely black, even from dark-sky locations. Therefore, do not adjust input levels in order to make it so, as you will only limit the number of usable gray values in the raw images. Sky background darkening is best achieved when making final adjustments in the finished stacked image.

Fig. 3.8. Recording and maintaining the maximum number of realistically possible discrete gray shades in an image can be the defining aesthetic difference to achieving a photorealistic appearance (NGC 1232).

3.10 Understanding Capture Rates

One of the common problems for newcomers to deep-sky video astronomy is, understanding how the frame capture rate of your capture interface relates to the accumulated image refresh rate of the camera. The benefit to understanding the fundamental difference will ensure you do not create overly large movie files containing repeated images. Furthermore, this will substantially reduce demands on computer processing and hard disk storage capacity. What we want to obtain is a movie file containing a string of unique images captured at a rate equal to the cameras accumulated image output refresh rate.

It is important to remember that no matter what mode the camera is set to, it is constantly outputting full frame pictures at a clocked rate of 1/25 or 1/30 of a second regardless of the exposure setting or frame accumulation setting. If you are doing occultation or planetary imaging, for example, you can set the video-capture rate of your recording software to any desired value from, say, 30 or 25 fps, down, to 1 fps, depending on the level of timing accuracy you require.

In the case of a 1 fps capture setting we will only be digitizing one of the 25 or 30 pictures streaming from the cameras output as a single image each second. As we mentioned earlier, this slowest typical capture rate will put the least demand on your computer's processor. Since changes in atmospheric seeing conditions are more often than not a rapid event, planetary imagers will use faster frame capture rates from, say, 15 fps to 25 or 30 fps, in order to capture the sharpest moments.

At the other end of the scale, frame accumulation mode uses the maximum possible camera exposure time of 1/50 or 1/60 of a second, and user selectable shutter control intervention is disengaged because the camera must now build up an image revealing the faintest detectable light it can achieve. If using the maximum accumulation mode setting, say X128 or X256, the faintest detail or object possible (within the limits of the camera and optics) now reveals itself on the video monitor. The actual period of time or delay between each unique accumulated image output by the camera is known as "picture refresh time" (PRT) and is directly related to the accumulation setting specified.

With each incremental accumulation setting below the camera's maximum, picture refresh rates occur with increased frequency. For example, X64 mode will produce a new unique accumulated image within about 1.28 s. See Table2.2 of accumulation rate versus screen refresh times in Chap. 2.

Therefore, using, say, the X128 accumulation mode of the GSTAR-EX, for example, a new or rather, refreshed co-added image will be the output by the camera approximately every 2.56 s. If you set your capture software to a rate of 25 fps, then your movie file would in fact contain 25 copies of the same accumulated image for each second of footage.

For example,

A 1-min recording containing unique images output every 2.5 s is (60 s/2.5 s capture rate = 24 unique images).

A 1-min recording captured at 1 fps equals (60 s/1 s capture rate = 60 images).

The latter capture rate yields a movie file that contains about 60% repeated images. This is especially poignant when it comes to image stacking, where its function to reduce noise is more efficiently achieved using only images with unique, random patterns of background noise. Therefore, a capture rate matching or closely matching the camera's output refresh time is best.

Since the slowest frame capture rate in commercial software is 1 fps, astronomy-specific programs, such as GSTAR Capture from Binary Systems P/L with later enhancement by Chris Wakeman, enables slower capture rates that can be set to match the frame accumulated image refresh time of the camera. This means the movie file produced contains a sequence of unique exposures and background noise so that when stacking a picture with, say, 300 captured frames, the resulting image is indeed a stack of 300 "unique" images.

Using our X128 accumulation rate example, if you captured, say, 300 images at 1 fps (a 5-min recording), the repeated frames do not realistically count in terms of noise reduction effectiveness. If you stack all these images, the result is the equivalent of 117 unique images. So to achieve the same result as a stack of 300 unique images using a capture rate of 1 fps you would need to record for about 12.5 min and stack the resulting 750 images, which includes all the repeated ones.

To calculate the recording time RT = PRT × UI/60 s, where RT is the recording time needed, PRT is the camera picture refresh time, and UI is the desired number of unique images required for later stacking.

The essential difference really comes down to the megabyte file size of the movie created and how many more images need to be processed in a stacking program. With most modern computers today having fast processors and large disk drives, these factors may seem a trivial matter, since a good result can be achieved either way. The main point being … use a capture rate closely matched to the image refresh output rate you have set the camera to where possible (Fig. 3.9).

Camera set to X128 integration, generates 1 unique frame every 2.56 seconds.

Recording device set to record at 25 frames per second
will record 63 uneeded duplicate frames.

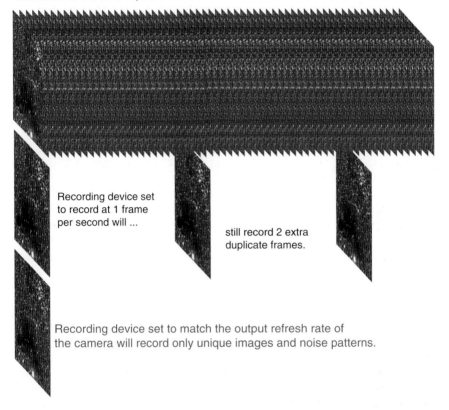

Recording device set
to record at 1 frame
per second will ...

still record 2 extra
duplicate frames.

Recording device set to match the output refresh rate of
the camera will record only unique images and noise patterns.

Fig. 3.9. The diagram here shows vividly the importance of matching the recording device capture rate to the camera's unique image output refresh rate to avoid producing unnecessary repeated images.

3.11 Storage Mediums

It is amazing to realize that some of the crudest records of human history left in stone wall paintings and rock carvings thousands of years ago still have longevity today. Yet, with all our current technology, the digitized records we now gather are only as good as the compatible devices capable of reading them in the future.

One great benefit of the digital age is being able to transfer or convert information with relative ease from one format to a newer one before the original recording/playback device has become redundant and unavailable. However, if a killer meteorite ever visits our planet and an ice age or water world follows, it may only be those ancient carvings of our ancestors that will still be decipherable by future generations long after the ice or water has subsided!

But our modern digital mediums, no matter how temporary, offer us a marvelous way of re-accessing and working with data, be it text or picture, and your deep-sky videos can be captured and stored in a number of ways, including magnetic tape mediums, such as video cassette tape, mini DV camcorders, or optical mediums such as CDs and DVDs, including their re-writable counterparts. You can also record to computer hard disk drives or RAM (random access memory), USB memory sticks, etc. And, the data contained within each of these storage mediums can be transferred from one to the other.

3.12 File Sizes

Since a movie file contains a number of multiple sequential images, it is naturally larger in terms of file size (bytes) than one single image extracted from it. For example, a single uncompressed frame taken from a 768 × 576 pixel AVI video file is around 1.24 Mb. Sixty of these uncompressed images in a movie file is 60 × 1.24 = 74 Mb, so you could only store around ten of these movies on a standard single-layer

Fig. 3.10. A variety of storage mediums, including analog and digital formats.

CD-ROM or 113 on a 8.5 Gb DVD-RW. The increased storage capacities of USB memory sticks make them a very popular and compact re-writable option these days. Although most computer hard disks nowadays have enormous storage capacity, it seems we tend to clog them up quickly with rarely used programs and other unessential data, so be sure to check your available disk space before venturing into a long night of video capture (Fig. 3.10).(Fig. 2.7)

CHAPTER FOUR

At the Telescope

Before we venture off into the practical side of a typical video imaging session under the stars, it is worth mentioning just a few of the commonly employed camera accessories and configurations you may find useful. Some configurations may require the use of threaded step rings or other adapters to allow for successful attachment.

4.1 Focal Reducers

Focal reducers are great for producing a wider field of view, especially when using cameras with small CCD image sensors. They essentially reduce your telescope's focal length so that it performs like an instrument with faster optics of the same aperture, yielding brighter images. Some can introduce coma (elongated stars appearing around the edges of the field of view) and may require a coma correcting lens also to ensure a flat field of stars across the entire image.

Another point of note is that certain telescopes with fast f/ratios like Newtonians and some short tube refractors may experience inward focusing problems when used with a focal reducer. Check with your optical dealer or browse the discussion forums for comments on your particular telescope combined with the focal reducer you might be considering. Depending on the design, often the reducing lens must be situated very near the camera's CCD for yielding the flattest field of view or simply just to achieve focus. C-mount reducers, like the Vixen 0.6X, are very good at allowing you to screw the camera directly to the C-mount thread at one end and a male T-thread at the other. Coupled to a low-profile 2-inch to T-thread adaptor like that manufactured by Lumicon, it may be used with most telescope designs (Fig. 4.2).

S. Massey and S. Quirk, *Deep-Sky Video Astronomy*,
DOI: 10.1007/978-0-387-87612-2_4, © Springer Science + Business Media, LLC 2009

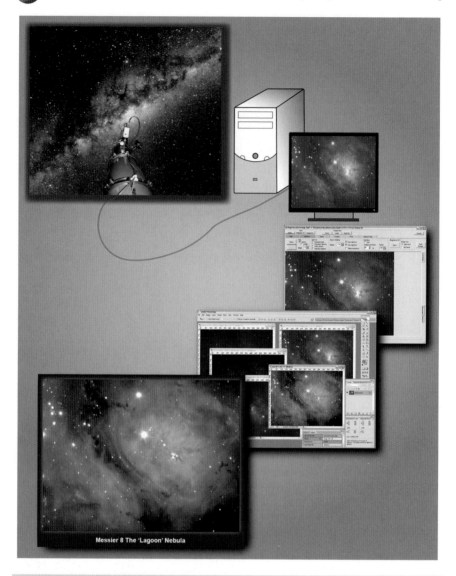

Fig. 4.1. From sky to portrait.

Messier 8 The 'Lagoon' Nebula

4.2 Barlow Lens

Mainly used for planetary work, a Barlow or tele-negative lens will increase image scale or effective magnification of the subject being recorded. They can be found with magnifying steps of 1.5×, 2×, 3×, 4×, and 5×. They are very useful for producing scaled up views of small galaxies or planetary nebula with high surface brightness, but be aware that images will become proportionally dimmer with increased magnification.

Fig. 4.2. A C11 Schmidt Cassegrain fitted with a Meade f/3.3 focal reducer via camera adaptors. Because of limited focus travel, this configuration does not allow most common filter selectors to be fitted into the optical train. Courtesy of Mike Holliday.

Fig. 4.3. Vixen 2.4X C-mount Tele-Extender nosepiece model # 3748. Ideal for increasing image scale with C-mount video cameras.

Vixen Japan manufactures a very useful 2.4X C-mount lens that screws directly to the body of most security surveillance-based cameras. It acts not only as a Barlow magnifying lens but also as a standard 1.25-inch nosepiece that allows you to place the camera directly into a standard telescope focuser with the added advantage of minimizing flexure in the optical train (Fig. 4.3).

4.3 Eyepiece Projection

If you do not have a Barlow or only need a little more magnification than your existing tele-negative lens can produce, then eyepiece projection may be a suitable alternative, although in practice it is rarely used for deep-sky imaging. However, in best seeing conditions with, say, a fast Newtonian, highly luminous yet small objects, such as the Hour Glass Nebula at the heart of M8 or the Homunculus Nebula around Eta Carina, may be worthwhile imaging with this method, but a Barlow lens is certainly the preferred option. A basic projection adapter usually has a 1.25-inch nosepiece at one end and a T-thread for attaching a camera at the opposite end. A low-power eyepiece of your choice is simply inserted and fixed inside the unit, and the whole assembly is then fitted to the telescope focuser. The image is now focused via the eyepiece directly onto the camera's CCD. Adapting the camera will require a C-T thread adaptor (Fig. 4.4).

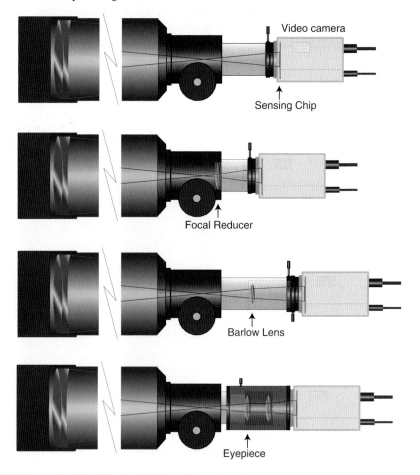

Fig. 4.4. Various optical configurations to help you produce a desired image scale from your camera and telescope.

4.4 Filters

Filters can enhance or reveal subject detail by allowing specific wavelengths of light to pass or be excluded. For example, the scattered artificial light in a suburban night sky can be excluded by a light pollution reduction (LPR) filter.

Narrow band filters are tuned to allow only a specific wavelength to pass. These narrow band filters are ideal for imaging specific emission and planetary nebulae.

Important for accurate tri-color imaging is the infrared-blocking filter (I-block). Since, nearly all CCDs are sensitive into the IR part of the spectrum, an I-block filter is used to essentially cut off this wavelength, thus retaining color integrity within the visual spectrum when imaging with RGB filters. Moreover, this filter will reduce the bloated appearance of stars caused by IR leakage. Baader Planetarium makes an ultraviolet- (UV-)/IR-blocking filter that includes the added benefit of also blocking wavelengths at the UV end of the spectrum, if needed. The downside of using an IR-blocking filter is that the image appears less bright. For general video viewing purposes an I-blocking filter is not essential (Figs. 4.5, 4.6 and 4.9).

Other filters worth experimenting with are the OIII, Deep Sky, Ha, and 1 micron. In areas of highly light polluted skies the Baader Neodymium (Moon and Skyglow) filter can be very useful for improving background sky contrast, and it also includes built-in UV/IR blocking (Fig. 4.8).

4.5 Filter Selectors

For efficient tri-color imaging, a filter selector is mandatory for maintaining correct camera orientation between each filtered exposure and quick interchanging. Your I-block filter should be fitted to the camera's nosepiece or at the focuser end of the

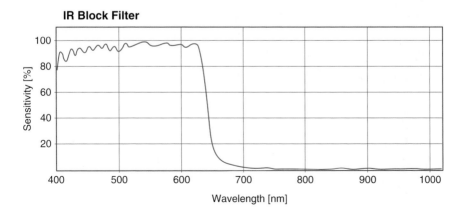

Fig. 4.5. Graph showing the spectral response effect of an IR blocking filter placed in front of a CCD.

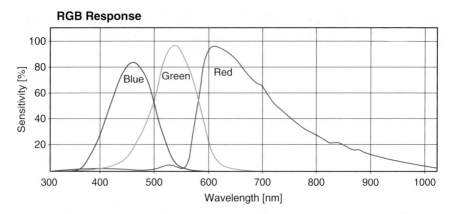

Fig. 4.6. Here we see the typical spectral response of a *CCD* chip at red, green, and blue wavelengths.

Fig. 4.7. Popular Lumicon filter selector fitted with *RGB* filter set.

filter selector so that each color filter is appropriately managed for IR leakage. One unfortunate aspect of filter selectors is that they add length to the optical train and might pose problems achieving enough inward travel in order to obtain correct focus at the camera. This is particularly the case with fast Newtonians, which may require replacement of the existing focuser to one with a lower profile or more drastically, moving the primary mirror forward. When using filter selectors with Cassegrains or refractors, achieving focus is generally not a problem (Fig. 4.7).

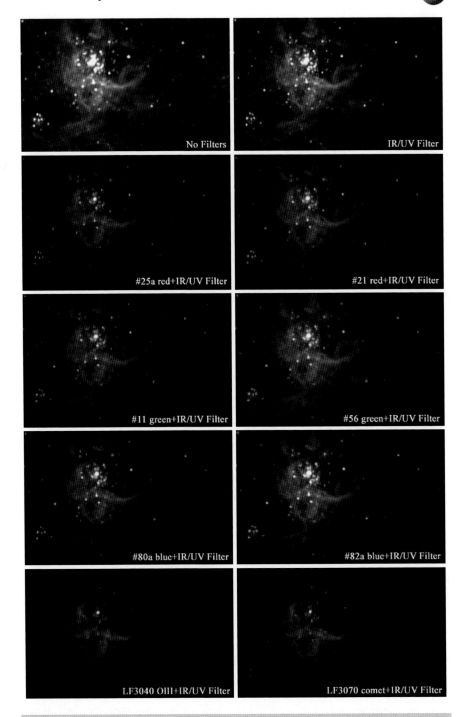

Fig. 4.8. NGC 2070 imaged with integrating video through a variety of optical filters.

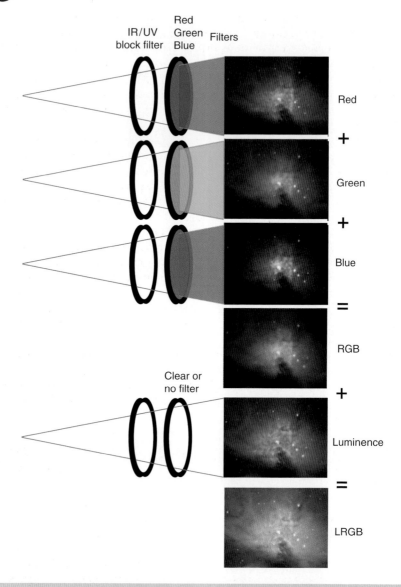

IR/UV
block filter

Red
Green
Blue

Filters

Red

+

Green

+

Blue

=

RGB

+

Clear or
no filter

Luminence

=

LRGB

Fig. 4.9. Recording an image with red, green, and blue filters produces an RGB image. By combining an unfiltered image called a luminance exposure, you can create an LRGB image.

4.6 Real-Time Video Observations

There are many amateurs who simply enjoy using a deep-sky video camera purely for observing because it can so readily reveal more than the eye can see. Even when fitted to a 100 mm *f*/9 refractor, the central star of the Ring Nebula is quite visible on a monitor from light polluted suburbs, so you can well imagine the joy of seeing a near 17th magnitude object with a 12-inch telescope from a dark country sky.

From your personal space at home, firing up bright monitors may be perfectly acceptable. However, at a star party gathering, where dark eye adaptation is so precious to purist visual observers, you will soon find yourself the subject of verbal abuse! So to maintain dark site etiquette you should do your best to situate yourself as far away from others as possible. Some star parties these days even have dedicated areas for digital astrophotographers. Even from a distance, the glow from your monitor may be visible to others, so set yourself up behind a car or tent opposite the main observing area. To further minimize extraneous light, it is worthwhile constructing a simple cardboard box housing with transparent cellophane or a perspex red screen. In fact, you may notice that it will also improve visual contrast between object and background sky. This makeshift housing will also protect your valuable laptop or CRT from dew, which might potentially cause damage.

Since portrait making and, therefore, later stacking images is not the objective of the "point and see" videographer, spotting a really faint object on the monitor alone is confined to camera and video display performance alone. If the camera is set to optimal maximum sensitivity then simple adjustment of the brightness and, particularly, the contrast control of the monitor can mean the difference in confirming the faint diffuse nature of a suspected comet or supernova in a far off galaxy. Monochrome CRT displays like those used in the security industry are highly recommended for best visual contrast.

And speaking of all this gear, powering it all is an obvious requirement when away from your home base. So, ensure you have an adequate battery supply to power your telescope, camera, and computer throughout most of the night, and keep video and power cables neatly tucked out of the way to avoid potential disasters when random observers pop over for a quick visit.

4.7 Creating a Video Portrait

If it is a nice picture you are after then you are probably chomping at the bit to get things started! But before diving into the deep end for that long awaited video portrait of your favorite galaxy, there are a couple of other helpful procedures that are well worth knowing about. There are necessary steps (additional images) you should undertake at the telescope as part of your astro-imaging session in order to correct camera or optical defects later when it comes time to process your masterpieces.

4.8 Dark Frames

Dark frame exposures are an important requirement for the removal of random and permanent hot pixels and also amplifier glow gradients present on the CCD. The dark frame is subtracted from the sky frame during post-processing to leave only the real information coming from the sky. In the case of hot pixels, it is vital they be removed so as not to present "false stars" in an image. A dark frame is made by covering the optics with a cap or cover to exclude any external light from the sky reaching the

CCD so that only the noise produced by the camera is recorded. Camera settings for taking dark frame exposures should be the same as those used for the sky exposure. The capture duration governing the number of images output by the camera should also match the number of sky frames captured. For example, if you are imaging a galaxy (the sky frame) for about 4 minutes and 16 seconds, then the dark frames must also be recorded for 4 minutes and 16 seconds. Though if you do long video recordings, for example 30 minutes or more (several hundred 128X sense-up frames), you may find it possible to get away with a smaller quantity, say 30–50%, of dark-frame recording. Ultimately, this will come down to your own trials (Figs. 4.10–4.12).

Depending on ambient temperature changes throughout the night, dark frame exposures should be taken either before or just after each sky exposure. However, if outside temperatures remain fairly constant over a long period and the pattern of visible hot pixels appears stable, then it is possible to use the same dark frame exposure for processing several images taken throughout the night or over a few hours.

The dark frame movie file is stacked to create a master dark frame image. The master frame has the random noise averaged out, so that only the thermal glow and hot pixels are left.

An unfortunate result of removing totally saturated (permanent) hot pixels is that a dark space will be left in the finished image. This is because the process needs a value of less than 255 (totally saturated) for the numbers to work on, for example, a value of 255 in the sky image minus a value of 255 in the dark frame equals a value of 0 (black). But all is not lost, as these dark spaces can be easily removed in the final processing, as we shall see later.

A solution to the hot pixel problem in warm weather conditions is quite obvious – keep the camera cooler. Cooling can be achieved with a simple flexible cold pack, like one used for sports injuries. Wrapping one around the camera will keep it cold for a few hours. A more permanent solution is to attach an electric cooler. Low cost coolers can be cobbled together from small refrigeration devices, such as car or drink coolers. Take out the peltier cooler, heat sink, and fan assembly and attach it to the camera. Power it from a 12-V battery or power pack. This will quickly cool the camera in several minutes but has no temperature regulation, so a set temperature cannot be maintained. Be aware that if the camera drops below the dew point, moisture could condense within the camera, and this could cause the internal circuitry to short out, thus ruining the camera. Indeed some suppliers are now adding or integrating small peltier cooling units into their cameras. However, these add-on cooling units will increase the cost of the camera considerably.

4.9 Flat Fields

Flat fields are a way to remove variations in the background of the image. These can be caused by differences in sensitivity across the CCD chip, vignetting of the optical path, or dust on optical windows and filters. Although they are not essentially needed to produce aesthetically pleasing images, we will cover the process here for those who may find it useful for a given application.

A flat field AVI recording can be done in the twilight sky when no stars are visible or using an artificial light source. A third method is to take a sky flat; this is a

Fig. 4.10. Sky frame of the subject shows some amplifier glow and several hot pixels (NGC 7582-90-99).

Fig. 4.11. Dark frame of the same recording time and settings as the sky frame.

Fig. 4.12. Dark frame subtracted from the sky frame.

median image of a multitude of sky frames. But typically video images will have a bright target in the center of the field, biasing the levels across the flat. This can be compensated for by turning off the drive and recording a large number of frames. For typical use, this becomes rather time consuming.

For greatest flexibility, construct a purpose-built light box that can be fitted over the telescope aperture or an illuminated screen in the dome or a simple white cloth (like a T-shirt) stretched over the telescope aperture and pointed at a light source. As long as the result is an evenly illuminated diffuse light, it will serve the purpose for creating a flat field.

Exposures must be made using exactly the same physical/mechanical camera configuration as used on the telescope for the sky frames so that the shadowy patterns are not altered. In other words, take flat field exposures before removing the camera after your imaging session is finished. Even changing focus or sliding in a new filter can alter these subtle patterns. By having an illuminated source you can take flats at your discretion during the night, and the results will be much more consistent. This then allows you to alter optical configurations at will and take new flat fields for each change.

The exposure for the flat frame should be a strong signal, that is, not noisy, so a reasonably bright target is needed. Adjust the exposure with accumulation mode "off," no gain, and a shutter speed set to make the image a medium gray level, that is, not too dark and not over saturated. The technical value is half the full-well capacity of the CCD. Capture about 100 frames at 25 fps to produce a stacked image free of random background noise.

To make a truly correct flat field image you must also take a flat field dark frame. The flat field dark frame must be taken using the same exposure levels as the flat field but with no light source. This flat field dark frame is later subtracted to produce a flat field with no thermal noise contribution (Figs. 4.13 and 4.14).

Fig. 4.13. Enhancing the left side of a flat field image reveals dust donuts.

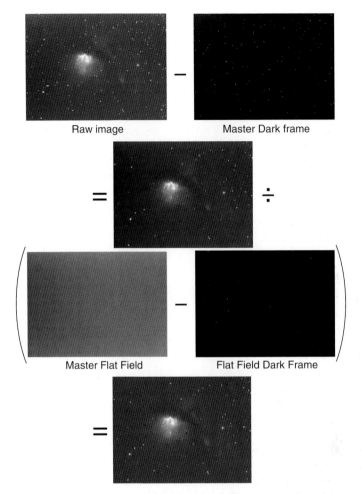

Fig. 4.14. Sequence of dark frame subtraction and flat fielding.

Great lengths and care must be taken to get good flat fields. For simplicity sake, most of the time we can get away with using only dark frames, but for the cleanest results, where assurance of complete artifact removal is mandatory, then flats and flat darks should be used. At the end of the night, for each optical configuration, you should have sets of recordings of sky frames of your deep-sky target, dark frames, flat frames, and flat-dark frames. How to utilize all these correction frames will be described in Chap. 5, Initial Processing.

4.10 Avoiding Dust Motes

As a general rule, always try to keep the chip facing downward when placing on adaptors and filters so dust will not settle on the CCD. Use of a clear or an IR-block filter left on the nosepiece will keep dust off the CCD's protective window. If you have to remove any dust, try blowing with a camera lens air blower. Compressed air tins or air compressors are useful but can sometimes spray contaminants onto the chip window. Use of lens cleaning fluids must be sparing, as you do not want excess fluid getting into the chip itself. Cotton tips or soft camel hair brushes will work well here. When cleaning, it helps to turn the camera on, remove any lens or nosepiece, and point it at the monitor or a bright light. This way, any dust will stand out clearly on the monitor.

Bias frames are the base electronic noise produced by chip circuitry. For the most part, it is not necessary to remove bias frames from video images unless processing is pushed well beyond normal limits. In practice, when a dark frame is removed, some portion of the bias noise will be suppressed as well.

Tip: During a big night of imaging, start by taking your first set of recordings followed by darks and flats; then, as you are acquiring and setting up on the next target, start stacking the previous set of AVIs. This way at the end of the night when feeling cold and tired, you will not have a long processing session still to get through.

4.11 Image Scale and Resolution

In the previous chapter, we discussed how a video signal is digitized and converted into a matrix of pixels to become an image and the relevant capture window sizes controlled by the capture device software drivers. However, recording the maximum detail possible within the theoretical diffraction limit of your telescope optics is governed by its aperture, the overall effective focal length used, and the size of the pixels in the CCD array of the camera. Since the finest detail resolvable with your telescope is determined by the size of the Airy disk pattern, the classic Nyquist sampling theory states that to reveal this fine image element effectively, it must span at least two pixels on the CCD.

First, we need to calculate the theoretical resolving power of the telescope. The popular Dawes limit calculation will give a rough idea being 4.56/(telescope aperture in inches). In the case of an 8-inch 2,000-mm focal length telescope, this gives us a theoretical resolution of 0.57 arc seconds in superb atmospheric seeing conditions. So we will need a suitable focal length to achieve 0.57/2 = 0.285 in order to sample this image element across two pixels.

Let us assume the camera we are using has a CCD with 10 μm pixels. By using the formula for calculating image scale in Chap. 1, we find 10/2,000 × 206 = 1.03 arc seconds. According to the Dawes limit calculation in this example, we are grossly under sampling our target and not achieving the potential image resolution. We must, therefore, increase the optical focal length in order to meet the Nyquist sampling requirement. A simple formula for determining a rough focal ratio required to satisfy the Nyquist requirement is to simply multiply the CCD pixel size by 3.5.

Since, it is generally the goal for most deep-sky videographers to simply record a nice portrait by fitting as much of the object on the video camera's CCD as possible, the need for conforming to the Nyquist sampling requirement seems pretty irrelevant and more within the realm of lunar and planetary imaging. But considering that such a high percentage of interesting celestial objects are typically quite small,

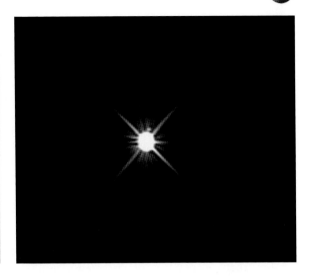

Fig. 4.15. Using the high power of a large telescope, the companion star of Sirius is resolved just 8 arc seconds from the bright primary. 20-inch reflector at f/5 (2,500 mm F.L.). Courtesy of Jonathan Bradshaw.

such as planetary nebula and many galaxies, meeting this criteria may prove quite useful in revealing fine detail otherwise unresolved using only shorter focal lengths (Fig. 4.15).

4.12 A Basic Plan for Your Imaging Session

Although most amateurs have their own individual setup procedures and checklist techniques, most newcomers experience the greatest difficulties in the field at video astronomy gatherings. For this reason we have provided the following "in the field" guide, which may assist in avoiding some frustration.

4.12.1 Step 1

Plan beforehand what it is you intend to image and how you want to record it. This might be a series of shorter accumulated exposures followed by longer ones. Consider where the object will be relative to the horizon for a given time. For example, if you are planning to do an LRGB image of an object rising from the east, then you should do the less important red, green, and blue exposures first followed by the luminance exposure. If the object is at the meridian or has passed through the meridian, then the opposite approach is best.

So why should you make your luminance exposure a priority? Well, from the perspective of the human eye, this is the best way to start. The retina is made up of photon receptors known as cones and rods. The cones that populate the central fovea of the retina make color perception possible and respond very well to bright light sources. Rods, on the contrary, are peripheral detectors, being more sensitive to low-light situations. They are, therefore, the most useful receptor when our eyes are dark adapted for observing faint deep-sky objects visually but do not mediate color response.

Since our eyes respond well to the luminance (black and white aspects) of an image, which includes all the primary visual wavelengths, the luminance exposure is subsequently brighter overall, yielding more extended detail than that seen in color filtered images. So to ensure the sharpest detail in an LRGB image, the luminance exposure should be taken when the object is at its highest elevation, where the atmosphere is thinnest and less affected by the distorting effects of air turbulence.

4.12.2 Step 2

Now that you have an observing and imaging plan, it is time to set up the scope. To ensure the best tracking, and, therefore, the least need for constant object recentering throughout the night, ensure that your mount is level and correctly set for your latitude and as accurately polar aligned as possible.

4.12.3 Step 3

Balance the telescope with your camera and any other accessories such as filter selector or focal reducers that you will be using. Keep interconnecting signal cables clear of mains power extension cords, and ensure all cables, including power leads, have adequate length to meet all possible telescope positions to avoid potential disasters. Remember that cables will also have an effect on telescope balance, so consider running them along the optical tube if possible, using adhesive cable clamps to help centralize overall weight distribution.

Carry out any two or three star alignments required in the case of a GOTO mount system. You can temporarily replace the camera with an eyepiece, although you can just as easily use the camera for centering stars during the alignment process.

Although not essential but often preferable, make sure the image as seen on the monitor is not reversed or mirrored for the particular optical system you are using. Newtonian telescopes will be correct, but telescopes such as Cassegrains and refractors will invert the image, and if a star diagonal is used it will be mirror reversed as well. Some cameras and even capture software programs provide functions to allow the image to be flipped appropriately. If not, you can of course achieve the correct orientation later when processing the final results in your favorite image editing program.

4.12.4 Step 4

Check brightness and contrast levels of your display monitor. There is nothing worse than having spent a night recording at the telescope only to find your movie files are too dark or oversaturated simply because the brightness and contrast settings of your preview monitor were poorly adjusted for correct visual interpretation of the camera's exposure setting. Even if you are just doing live viewing, correct monitor brightness and contrast can mean the difference between seeing a faint galaxy or not.

And if capturing to a computer, it is imperative to check capture driver settings for optimal brightness and contrast output so that grayscale levels are maximized. Some image-capture programs provide a snapshot histogram as a guide for ensuring you have optimal settings to obtain the widest range of grayscale in the image. As mentioned in the previous chapter, in this modern light-polluted world we live in, the background sky is more a dark gray than black, so do not adjust levels to force a perfectly black sky.

4.12.5 Step 5

After allowing ample time for the telescope optics to cool to ambient outside air temperatures, check the focus of the camera. Like single-shot cameras with LCD displays, focusing with video is a pretty simple process because you have a live feedback loop showing exactly how sharp the stars look on your monitor. With no frame accumulation set, any bright star in the field will be easy to focus on. If no bright stars are in the field, then set frame accumulation mode to around X6 or X8. You will then see fainter stars appear while still maintaining a rapid near real-time response to changes in the focus position. Even very bright stars can be focused on by making any diffraction spikes symmetrical. A handy tip for video cameras that feature a digital zoom function is to use this option to obtain really fine focus with the zoom set to maximum. Then, return to normal (no digital zoom) for imaging (Fig. 4.16).

If you are having trouble focusing all the stars across the field of view, it could be due to poor collimation of the optics, quite often accentuated in fast optical systems. An 8-inch *f*/4 Newtonian was used to take the images shown in Figs. 4.17 and 4.18. Focal reducers can be another cause of stars not focusing right across the field, so check that they are seated correctly.

Also always check to make sure that the camera is mounted exactly square-on to the focuser and the optical axis.

4.12.6 Step 6

If using an auto-guiding system, select your reference tracking star nearest the object you intend to image. If not, skip to Step 7.

4.12.7 Step 7

After turning on the camera and setting the desired frame accumulation setting, adjust the video capture program to the required frame capture rate best corresponding to the camera's picture refresh time. Select the video format, window capture size, and video compression algorithm. Remember, it is best not to use any image compression if possible, so as to maintain raw image data integrity.

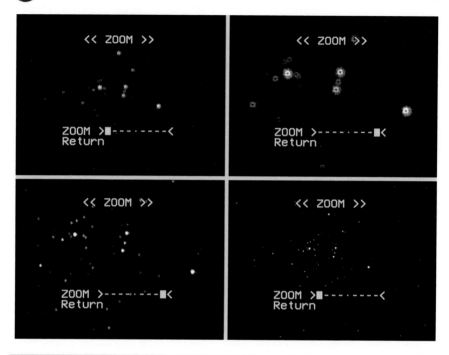

Fig. 4.16. Stars are out of focus; zoom in, focus the telescope, and zoom back out for imaging.

Fig. 4.17. A telescope with poor collimation will not focus correctly across the whole field of view.

Fig. 4.18. Same telescope as used for Fig. 4.17, but with correct collimation.

4.12.8 Step 8

It is often preferable to fix your camera into the focuser with an orientation of north at the top of the field of view. This is the way most astronomical images are presented. It also helps when moving the field by hand controller or dials so that stars move in a logical way across the screen. Placing a mark or a sticker on the focuser or telescope to indicate north will help each time you insert the camera.

However, this is not a strict rule and more a convenience, so consider the best camera angle needed for the object you are imaging. A classic example is the large galaxy NGC 253 at the focus of a 10-inch f/4.9 Newtonian. The camera is best oriented diagonally in this configuration to capture the full extent of its spiral arms from one corner of the monitor to the other. Adjust the camera by turning it within the telescope focuser, then tighten off the focuser thumb screws firmly.

At this point you can either cover the telescope and take dark frames or start the recording process and take dark frames at the end.

4.12.9 Tip

If doing RGB or LRGB exposures, be sure to note which AVI file relates to the filter that was used. This will help you later when it comes to processing and combining each filtered monochrome image. Naturally, this requirement is irrelevant in the case of a color camera.

4.13 Guiding and Tracking Corrections

If your telescope is not perfectly polar aligned or your motor drive and worm gears have some periodic error, then you might need to do some occasional recentering or rough guiding using the drive hand control unit. Fortunately, this is done quite easily by viewing drift and correction straight off the video monitor while recording your object at the same time. One simple method to help achieve a reasonably well-tracked object in such situations is to place a simple crosshair on the monitor.

This can be done by either placing two narrow strips of masking tape directly on the screen or making a cover screen from a piece of perspex with red cellophane behind it. The latter makes a great night screen, and you can mark crosshairs on the perspex with a whiteboard marker, which then easily rubs off later. An added bonus of the night screen is that it suppresses some of the random noise for live viewing. Another alternative is using a software-generated crosshair to place over a star with your mouse. This will also help you to monitor polar alignment star drift in order to facilitate any necessary minor adjustments to the mount (Figs. 4.19 and 4.20).

Fig. 4.19. Crosshairs marked on a night screen with whiteboard marker pen.

Fig. 4.20. Software-generated guiding reticle.

If for any reason you need to take the camera out of the telescope while making a series of images of a particular object, mark the position of a couple of stars on the night screen. Then, when you replace the camera, the field can be orientated correctly.

4.14 How Many Images Should be Captured?

To produce a reasonable final image with far less noise than the single image we see on the preview monitor, at least 100 unique video frames are needed, while 200 will show a noticeable improvement in noise reduction and increased image detail when stacked later. To make further obvious improvement, you can double the number of stacked images again, such as 400 frames, then 800 frames.

At around 800 frames we reach a point of diminished returns, so to make further significant improvements requires a lot more video frames for a marginal return. But do not let that stop you if you are willing to record for that amount of time. In a test, we took 1,600 images in lots of 100 and then stacked them in a series; major improvements were quite obvious up to 800 frames. There was a slight but still noticeable difference at 1,200 frames, which is only 50% more than the 800-frame stack. But it was clear that at these larger stacked image numbers, the 8-bit image depth was not revealing any more subtle gray levels once the overall noise had been well and truly smoothed out. In our experience, the happy medium particularly for the luminance exposure is either 400 or 800 frames for most deep-sky objects to produce a stacked image with good signal-to-noise characteristics (Fig. 4.21).

So, when recording video images to your computer, should we capture all our 400 frames in a single large AVI file or in batches of bite-sized video files containing, say, 100 images in each? Well, this largely depends on your own preference and method of recording. In the case of capturing, say, 4 × 100 images (4 recordings

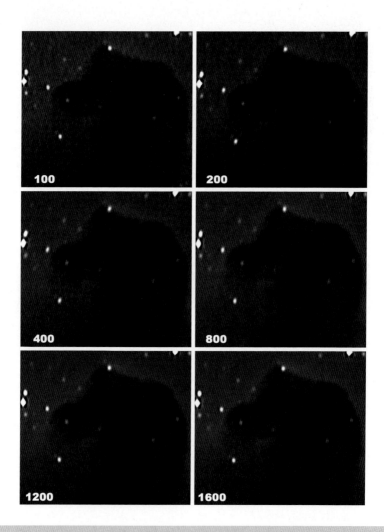

Fig. 4.21. Comparison of a 100, 200, 400, 800, 1,200, and 1,600 video image stacks.

of 100 video frames) and using a program such as Registax to import the AVIs directly for processing, then the batch method has its benefits in that only a 1 × 100 dark frame exposure is needed to process all four sky frames. The resulting four images can be stacked using Registax or the procedure described in Ch 5.3. Although you will repeat the dark frame subtraction and stacking process four times over, the final result can be achieved in less time than waiting for 400 dark frame exposures to be completed and then processed with the much larger sky frame AVI. But this naturally leads to more individual files to be managed on your hard drive, that is, five video and five resulting images (ten files) versus two video and two images (four files).

There is always the option of having all exposures contained within one neat AVI movie file by simply placing the cap over the telescope immediately after the required number of sky frame exposures is complete. With some manual intervention using AVI management software such as VirtualDub, you can later split out the sky and dark frames as individual exported image sets or as two separate AVIs.

Another incidental noise-averaging method is a phenomenon known as dithering. In the natural occurrence of mount periodic error or slightly less than perfect polar alignment (where an object moves subtly around the image sensor), the light falls on surrounding pixels near to the ideal tracked pixel position. When stacking all the frames later, this helps to further smooth out noise. So if your mount tracking is not perfect, then do not worry too much, as it is helping in its own small way. However, this really only works with video cameras where many short exposures are being taken.

The advantage of dithering is also utilized when taking multiple images to search for asteroids or comets. This way a true moving object will make a linear motion against the background stars, whereas noise or artifacts will be nonlinear due to the scatter of stars, etc., on the imaging chip.

4.15 Framing Multiple Fields to Capture Large Objects

How frustrating it is when you hope to fit an object in its entirety into the picture only to find it simply will not fit! Objects with large angular sizes, such as Messier 42 (about 40 arc minutes) and NGC 253 (20 arc minutes), can represent a problem in terms of fitting onto the small image sensors typical of video cameras. Most commonly employed for this purpose are focal reducers, but if you want to maintain prime focus resolution, then taking multiple fields and stitching them together later will provide a larger image with the detail you want. All this requires some post-processing labor, but the practice is common among astrophotographers using mediums other than video.

To do this, we start by taking two fields and practice the image-merging process. When framing each area of sky, try to achieve a 10–20% overlap so the merging process will appear seamless in the final result. Some imaging programs have auto-merging capabilities to help achieve this. See the post image-processing chapter for a more detailed account of the steps and techniques involved (Figs. 4.22 and 4.23).

Fig. 4.22. Messier 42, 80 mm f/5.6 refractor (two fields merged). Courtesy of Darrin Nitschke.

Fig. 4.23. NGC 3372, 20 cm f/5 reflector (four fields merged).

CHAPTER FIVE

Initial Processing

Procedures, procedures! Life is full of them! But without clear, step-by-step instructions you might not be able to assemble that DIY corner cabinet or new guest bed without a whole lot of cursing and carrying on within the 10 min set-up time quoted in the instructions.

So in the following two chapters we present, in very much a step-by-step format, the most common problems videographers have when it comes to processing their recordings on a computer. There is no need to read the following word for word. You can simply refer back to any one of the processes when you need to. They can help set you on the path to producing really nice deep-sky portraits. After all, this is where the magic really happens!

5.1 Software for Initial Processing of Video Files

After capturing our deep-sky video to a computer we can begin the initial task of selecting the best frames and removing unwanted artifacts. This was once a laborious manual task many years ago, but now several excellent software programs allow us to do this in only a fraction of the time.

To produce a nice single image with greatly reduced noise, each of the individual images contained within the video movie must be extracted and then stacked. However, before the stacking process is carried out, each video frame in the movie should be assessed for consistency and quality. In other words, images that deviate from the optimum observing conditions should be discarded, for example, a meteor or an

S. Massey and S. Quirk, *Deep-Sky Video Astronomy*,
DOI: 10.1007/978-0-387-87612-2_5, © Springer Science + Business Media, LLC 2009

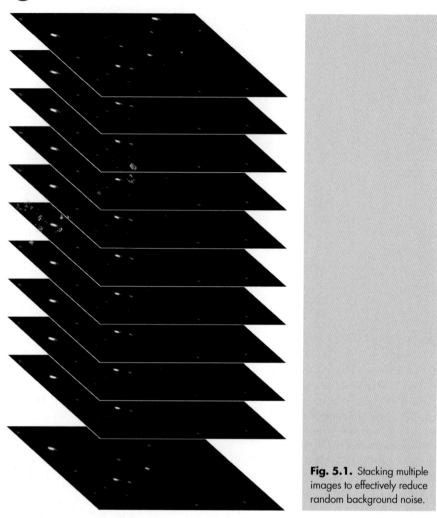

Fig. 5.1. Stacking multiple images to effectively reduce random background noise.

artificial satellite passing through the field, random patches of cloud dimming the subject from time to time, or a major shift in the mount caused by strong wind or inadvertent bumps to the telescope causing elongated stars. Even radical changes in the seeing causing the image to appear poorly focused momentarily should be discarded (Fig. 5.1).

There are several useful software tools available for reviewing and extracting individual images from a video movie. VirtualDub is one such freeware program that allows the user to move through a video file frame-by-frame using the arrow keys of your keyboard. If a particularly undesirable frame becomes apparent, one simply presses the delete key to mark it for removal. VirtualDub provides the option of saving the remaining video frames as a sequence of individual images. These images can then be imported into stacking programs such as the popular freeware "Registax" and others, such as AstroStack, K3CCDTools, or DeepSkyStacker, to name but a few that provide very efficient processing.

However, if you want to avoid the extra step of using VirtualDub, as it is a personal-choice application, the aforementioned stacking programs also include imported video frame selection tools allowing the user to check and uncheck specific images. Once this has been done, the image stacking process can commence and the program will use only those images with a checkmark against them.

Other programs, including DeepSkyStacker, can also align slightly rotated images to compensate for field rotation, such as those taken using Alt-Azimuth-mounted telescopes. Once the video frames are stacked they can be saved in the preferred image format. Although BitMap (BMP) images are quite often used, working with tagged image file (TIF) or flexible image transport system (FITS) files tends to yield the best results when it comes to processing the final stacked result. During the stacking phase, dark frames and flat fields (if applicable) can be applied to the sky frames to produce a final master stacked image, devoid of optical and thermal noise-related artifacts. At the time of writing, Registax (Version 4) is one of the most commonly used programs among amateur astronomers for carrying out this task and is used in the following descriptions and examples.

Depending on camera specifications, the video file created will contain either interlaced or progressive scan images. Interlaced video images often require post-filtering to bring the scan lines of each field together using a deinterlacing filter. Some capture software tools can do this on the fly before writing each image to a movie file. But if your capture software does not include a real-time deinterlacing filter (sometimes called progressive mode), programs such as VirtualDub can do this later. Registax also includes a deinterlacing filter. It can be set in the [General Options] tab above the main working area. Simply tick off the [Interlaced] checkbox.

5.2 Stacking with Registax Direct from an AVI File

If you have recorded dark frame and flat field AVIs, then you need to process these into individual images before stacking any deep-sky (sky frame) recordings (Table 5.1).

Of course, you can name the files anything you want, but it is best to keep it all logical, especially where many images are involved and you may only have a chance

Table 5.1. For the following descriptions we will use these file names.

sky.avi	Recording of the deep-sky object
dark.avi	Aperture covered, same recording settings and length as sky.avi
flat.avi	100 frames of illuminated diffuse light source
flat_dark.avi	Same settings as flat.avi with no light source
flat-flat_dark.bmp	Flat field image with flat_dark image subtracted

to revisit them some weeks after your imaging session when, of course, it all seemed very logical at the time!

When taking multiple darks and flats throughout the night, include your local or universal time (U.T.) in the file name as well, that is, "dark-2135," "flat-0245." If you archive the dark and flat frames, then put the date in the file name as well.

So, let us start by creating the master flat field. Note: If you are only using dark frames, then simply skip this section and go to 5.2.1.1 Creating the Master Dark frame and ignore further references to flat fields.

5.2.1 Procedure

To create the master flat field, do the following:

(a) From the [Select] menu, open the flat_dark.avi (Fig. 5.2).
(b) From the [Flat/Dark] menu select [Create Dark frame] (Fig. 5.3).
(c) Once this has finished, click the [Save Image] button and save as a BMP (default format).
(d) In the [Flat/Dark] menu select [Load Dark frame] and choose the flat_dark. bmp you just created (Fig. 5.3).
(e) Go back to the [Select] menu and open the flat.avi.
(f) Under the [General Options] tab, make sure the [Dark frame] checkbox is ticked (Fig. 5.4).
(g) From the [Flat/Dark] menu, select [Create Flat field] (Fig. 5.3).

This will now produce a single image of the flat field with the flat_dark subtracted. Save this image as a BMP called "flat-flat_dark.bmp."

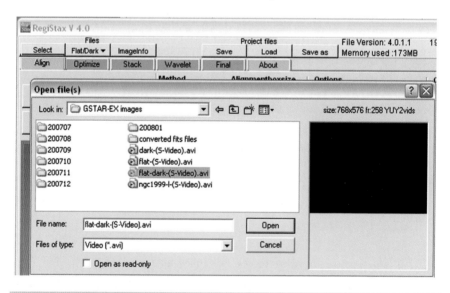

Fig. 5.2. Open an AVI dialog box.

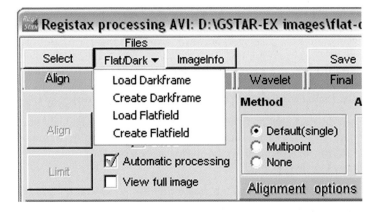

Fig. 5.3. Create dark frame or flat field menus.

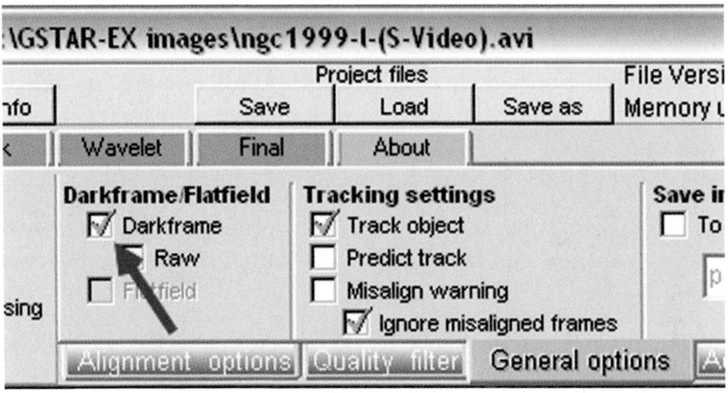

Fig. 5.4. Make sure the dark frame checkbox is ticked.

5.2.1.1 Creating the Master Dark Frame

(a) From the [Select] menu, open the dark.avi.
(b) In the [Flat/Dark] menu select [Create Dark frame] (Fig. 5.3).
(c) Once this has finished, click the [Save Image] button and save as a BMP (default format).

5.2.1.2 Loading the Master Dark Frame and Master Flat Field

(a) In the [Flat/Dark] menu, select [Load Dark frame] and choose the dark.bmp you just created.
(b) In the [Flat/Dark] menu, select [Load Flat field] and choose the "flat-flat_dark.bmp."

When you have multiple darks and flats, choose the appropriate ones for the corresponding time the sky.avi was taken.

5.2.1.3 Frame Selection and Stacking

(a) From the [Select menu], open the deep-sky target video, the sky.avi.

Here you can deselect any unwanted frames, such as those affected by poor seeing, distorted stars, meteors, or artificial satellites. Uncheck from the frame bar below the main working area by pressing the space bar or from the [Alignment Options/ Options/Show Frame List] box.

(b) Under the [General Options] tab, make sure the [Dark frame] and [Flat field] (if used) checkboxes are ticked (Fig. 5.5).
(c) Select your reference star from an active image. Be careful that you do not pick a hot pixel by mistake.
(d) Set the [Quality Estimator/Lowest Quality] to 0% so that all the frames are used.

Alternatively, if you want Registax to clip the lower quality frames, set the value to between 70% and 90% (Fig. 5.6).

(e) To save any further button clicking, tick the [Automatic Processing] checkbox. Alternatively, uncheck if you wish to make manual adjustments to the process as it goes along (Fig. 5.7).
(f) Click the [Align] button to start the processing sequence.

When the stacking process has finished, you have options to adjust sharpness (wavelets), brightness, contrast, and many other functions. Or, just save the raw stacked image.

(g) For this example, click the [Save Image] button and save as a TIFF file for further processing in Photoshop (Figs. 5.8 and 5.9).

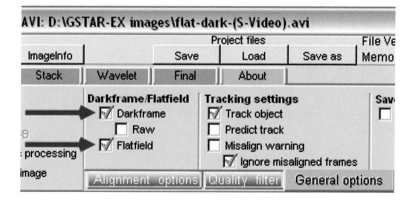

Fig. 5.5. Dark frame and flat field checkboxes ticked.

Initial Processing

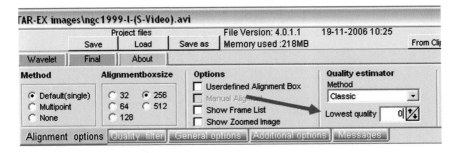

Fig. 5.6. Set the lowest quality option.

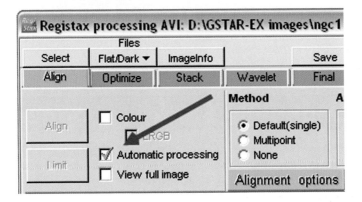

Fig. 5.7. Go auto.

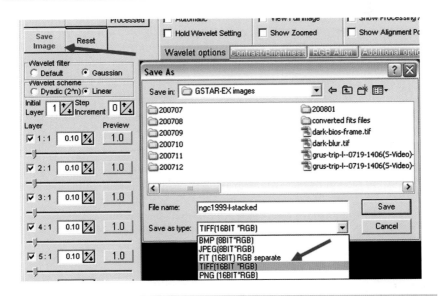

Fig. 5.8. Save the stacked image in preferred *TIFF*.

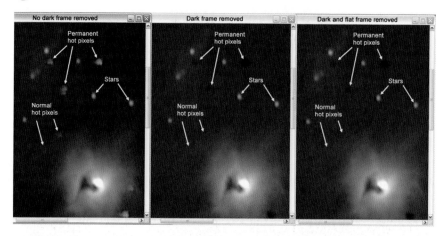

Fig. 5.9. Raw image, minus dark frame, divided by flat field (NGC 1999).

5.3 Manually Stacking Images

Every now and then and for one reason or another, certain images will not align and stack into correct registration. In these instances they can be stacked manually. It is rather a laborious task but definitely an option in cases of pure desperation.

There are two methods for doing this, using programs that provide image layering such as Photoshop.

1. If you have 2, 4, or 8 images, these can be stacked in pairs with a transparency/opacity blending of 50% for each pair combined. For example, 8 becomes 4 pairs, 4 becomes 2 pairs, and, eventually, 2 becomes 1, being the final stacked result.
2. A number of images can be aligned and stacked with a first reference frame. The following opacity percentages for each layer serve as a rough guide (100 divided by the number of frames) (Table 5.2).

In reality, a practical limit for the number of images using this manual procedure will be about 20 frames or so. To demonstrate manual stacking, we will use Adobe PhotoShop and the pairing method.

(a) Open the images to be stacked (Fig. 5.10).
(b) Select the first image, press [Ctrl+a] to select all, then [Ctrl+c] to copy.
(c) Select the second image, and press [Ctrl+v] to paste the first image as a new layer.
(d) Check the alignment by selecting from the layers palette, the [Difference] blend mode. Select the [Move Tool], click back on the image, and adjust with the cursor keys until aligned (Fig. 5.11).
(e) In the layers palette, set the blending mode back to [Normal] and the [Opacity] to 50%. From the layer menu, select [Flatten Image] and close the first image (Fig. 5.12).

Table 5.2. Layer stacking percentages.

2nd frame opacity at 50%
3rd frame opacity at 33%
4th frame opacity at 25%
5th frame opacity at 20%
6th frame opacity at 17%
7th frame opacity at 14%, and so on

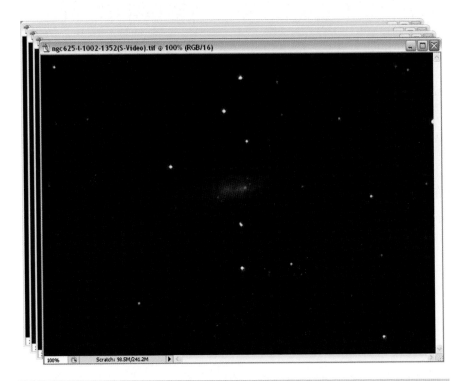

Fig. 5.10. Four images for manual stacking (NGC 625).

Repeat the same sequence for the third image onto the fourth image, [Opacity] to 50%, [Flatten] the fourth image, and close the third image. Now do all the same for the second image onto the fourth image. (There is no need to save the second image.) You are left with one image that has substantially reduced background noise (Fig. 5.13).

Save this image as a TIFF file, denoting it as a stack of four images to differentiate it from the original, that is, "NGC625-Lx4.tif."

Fig. 5.11. Adjust the images into register.

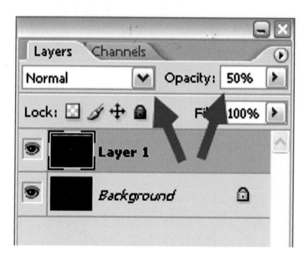

Fig. 5.12. Set the blend mode to normal and 50% opacity.

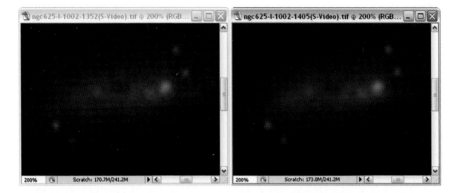

Fig. 5.13. Section of stacked image with noise now reduced.

No matter which process you use (hopefully the easiest), the outcome should be a single image of greatly smoothed appearance and dynamic range, magically created from all those noisy frames contained in the original captured movie file. This should now become your base platform for even greater enhancement using the steps outlined in the next chapter.

CHAPTER SIX

Final Processing

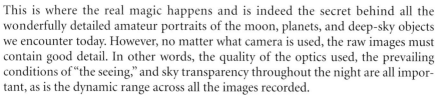

This is where the real magic happens and is indeed the secret behind all the wonderfully detailed amateur portraits of the moon, planets, and deep-sky objects we encounter today. However, no matter what camera is used, the raw images must contain good detail. In other words, the quality of the optics used, the prevailing conditions of "the seeing," and sky transparency throughout the night are all important, as is the dynamic range across all the images recorded.

In terms of our deep-sky video, as with any other form of astro-imaging, to get the most information from the final stacked result, a fine balancing act of refining overall brightness, contrast, color, and tonal qualities is more often than not a mandatory requirement and self taught skill. Other image manipulation techniques such as the "unsharp mask" can also be employed to extract fine detail out of your images to produce a truly great result.

Among the plethora of software for adjusting and enhancing images out there, tools such as MaximDL, Paint Shop Pro, AIP4WIN, Astro-art, Ulead's Image Editor, Images Plus, along with many others, are some good all-round tools. An open-source freeware program available over the Internet called GNU Image Manipulation Program (GIMP) boasts many features usually only found in commercial software. Astronomy-related programs such as MaximDL, also include many tools and several nice filters specifically designed for analysis and enhancement of astro-images.

The preferred software among most amateur astronomers is Adobe Photoshop, and therefore the following routines described in this section are based on this program. However, the principles applied for a given specific outcome are still applicable in practice where other programs provide similar functionality.

Photoshop has many powerful functions and is perhaps one of the best programs for manipulating multiple layer images. In particular, it is highly effective when carrying out the practice of layered unsharp masking. Additional plug-ins, such as

S. Massey and S. Quirk, *Deep-Sky Video Astronomy*,
DOI: 10.1007/978-0-387-87612-2_6, © Springer Science + Business Media, LLC 2009

Adobe Astronomy Tools, which have tailored macros that can improve and enhance your astronomical images, are also available.

Here are some quick PhotoShop tips and shortcuts that are very handy.

1. Throughout the following procedures where a palette is called for but not visible on the screen, it can be activated from the [Window] menu.
2. To make an image "active," simply click on its title bar.
3. Double click anywhere in the Photoshop work area to bring up the "open" dialog box.
4. When working with an image that is larger than the viewing window (or when zoomed right in), regardless of what tool you are using, hold the space bar down, and you will be able to move the image around in the window to see different areas. Release the space bar and the currently selected tool is again available.
5. When performing often repeated tasks, for example flattening layers, it is a great timesaver to create an "Action" button macro to execute numerous commands with a single click or keyboard press.

6.1 Calibrating Your Computer Monitor

An important pre-check step before opening and examining images on the computer is to check that your monitor is correctly calibrated. This can be achieved using special hardware tools available from some computer specialist suppliers. Most commonly used is the Adobe Gamma calibration tool that comes with the Photoshop package.

Grayscale bars or other test pattern images can be found on the Internet and are sometimes inherent to astronomy specific capture tools like the GSTAR capture program. A simple grayscale bar will aid you in discerning different shades of gray on the screen. By simply adjusting the brightness and contrast buttons on the monitor, you can maximize the number of shades seen from pure white through to pure black (Fig. 6.1).

You should only carry out this task in a well-darkened room as brightly lit environments will result in over compensation. During daylight hours, close the curtains and doors. In fact, when processing your images later you should also maintain a low-lit working environment.

LCD displays with low contrast ratios (less than 500:1) and high brightness levels [500 nits (candela/metre2)], typically found on laptop computers, will produce overly bright views of images that would otherwise look quite normal on a CRT monitor. The best LCD monitors available nowadays for image work should, therefore, have contrast ratios in the order of 2,500:1 or more and be capable of brightness levels down to around 100 nits. These will render your images with good tonal ranges similar to a CRT monitor.

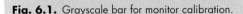

Fig. 6.1. Grayscale bar for monitor calibration.

6.2 Extracting the Detail

Perhaps the most exciting transformation to our stacked though not yet fully processed deep-sky image is the stretching of the all-important middle grays. When
applied it reveals an amazing depth of often invisible, faint detail that suddenly leaps
out of the image before your eyes, and its importance cannot be overstated enough.

By viewing the histogram of an image (a feature incorporated in Photoshop's
level adjustment tool) the distribution of light across the entire image or any selected
region is seen in graph form representing levels of pixel intensity.

Depending on the particular object in the image relative to image scale, its unique
surface brightness characteristics and how well the capture input adjustment levels
were originally set, the shape of the histogram may vary from quite a narrow profile
with less dynamic range to one that spans the full width of the range of discrete
gray pixel values. The wider the curve, the more that can be teased out of the image.
A good practice is to make a duplicate before each major adjustment, in case it all
goes horribly wrong or simply use the [History] feature to undo certain steps.

6.2.1 Procedure

(a) Open your image. From the [Image – Adjustments] menu, select the [Levels…]
    dialog box. Note how the histogram looks. If the histogram is spread out over
    the width of the graph box then only slight tweaking is needed. If the histogram
    is narrow in profile then it will need more tweaking (Figs. 6.2 and 6.3).
(b) Start by moving the right [input level slider] to the left until it gets close to the
    rising edge of the histogram.
(c) The left [input level slider] can then be moved slightly to the right to darken the
    background.
(d) The center [input level slider] can be adjusted to the left, increasing the faint
    mid-tone detail without revealing too much noise (Fig. 6.4).

Slightly tweaking of each of these sliders is often needed to extract the last bits of
information contained in the image. A little experimentation is always worthwhile
(Fig. 6.5).

Since faint targets will often produce images with a narrow profile or low dynamic
range, the same treatment must still be applied in order to bring it to some level of
prominence with reference to the background sky.

Fig. 6.2. Original image of NGC 3372.

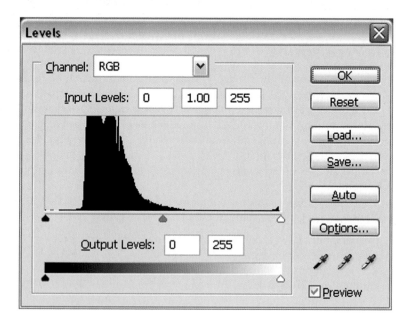

Fig. 6.3. Original levels histogram.

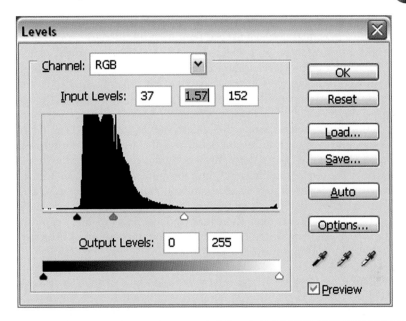

Fig. 6.4. Adjusting the input levels sliders …

Fig. 6.5. … will reveal much more detail in the nebula.

Again this requires moving the right [input level slider] to the left rising edge of the histogram to increase brightness, then moving the darkening adjustment to the right, pushing those pixel values toward the black end of the scale. Then adjust the center [input level slider] as previously described to highlight the middle gray regions. This is where practice and personal preference comes into play, so just experiment with each setting until a happy medium is reached (Figs. 6.6–6.9).

When adjusting the right [input level slider], moving too far left could bloat the brighter stars and even oversaturate brighter parts of a nebula, star cluster, or galaxy, resulting in a loss of this critical detail. By backing this control off to the right a little and adjusting the center middle gray [input level slider] a little more to the left, a more balanced and pleasing appearance can be obtained. In some cases, oversaturated areas are unavoidable but can be managed using other more advanced techniques, which we will cover later. After making these adjustments, open the levels dialog box again, and you will notice how the histogram profile has been stretched out to better fill the range of values from 0 to 255.

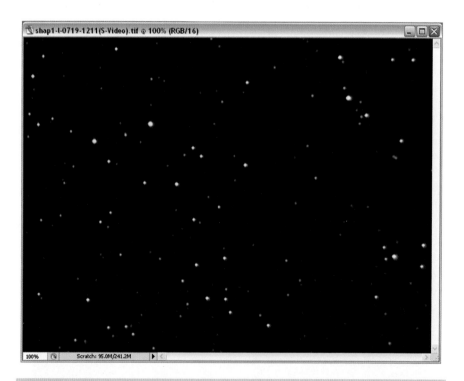

Fig. 6.6. Original image of Shapley 1.

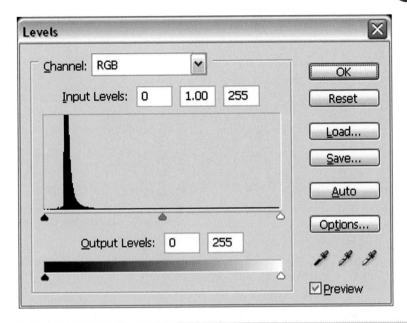

Fig. 6.7. This levels histogram shows a much narrower profile.

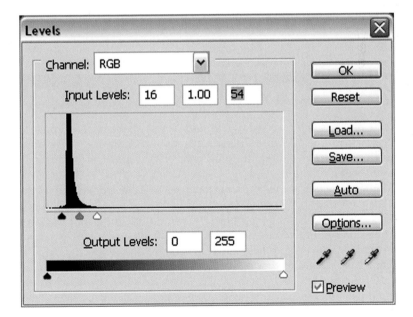

Fig. 6.8. Adjusting the input levels sliders to the histogram …

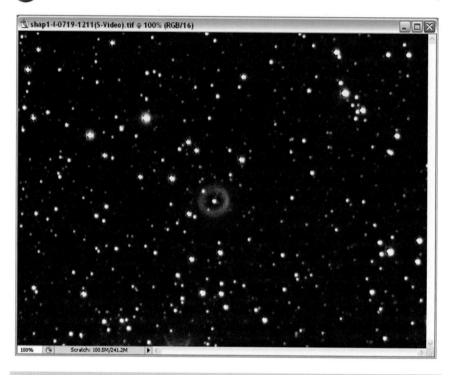

Fig. 6.9. ...will reveal the faint planetary nebula.

6.3 Vignette/Gradient Removal

Depending on the camera, optical system, and local sky glow, images produced may require some form of background gradient removal. These are effects caused by things such as vignetting or amplifier glow gradients generated by the camera's electronics, for example.

Pushing pixel values to the limits can often reveal faint amplifier glow that has eluded the dark frame processing. Normally, with smaller CCDs typical of video cameras, vignetting will not be a problem when imaging through a telescope, but with the addition of focal reducers or using other fast lenses, it might overexpose the sky glow more in the center of field, thus emphasizing a vignette.

To remove the offending glow and achieve a consistent background, you need to subtract this unwanted data from the real sky data. How this procedure works is by making a duplicate of the original image, then removing all the real data, that is, the object of interest and stars, using the Photoshop clone tool. What remains is just the unwanted glow. The glow image is then digitally subtracted from the original image, leaving only the real data recorded from the sky.

6.3.1 Procedure

The image used for this demonstration has the levels stretched too far, and a correctly matched dark frame was not used in the subtraction process; therefore, it reveals some undesirable camera amplifier glow from the upper left corner.

(a) Open the image to be processed.
(b) From the [Image] menu, select [Duplicate]. (No need to alter the duplicate name generated, just click OK.) (Fig. 6.10).
(c) Using the [Clone Stamp Tool] (set the brush size to about 20), remove any bright stars or extended objects from the duplicate image, making sure not to change the background gradient, as this is what we are trying to remove. Always sample the cloning point parallel to the glow gradient. Fainter stars will not affect the procedure once the image is blurred correctly (Figs. 6.11– 6.12).
(d) With the duplicate image still active, from the [Filter] menu, start the [Blur – Gaussian Blur ...] dialog box. Adjust the pixel radius so that there are no lumpy bits left and you see a smooth gradient. This image used a value of about 25 (Fig. 6.13).
(e) Now make the original image active, and from the [Image] menu start the [Apply Image] dialog box (Fig. 6.14).

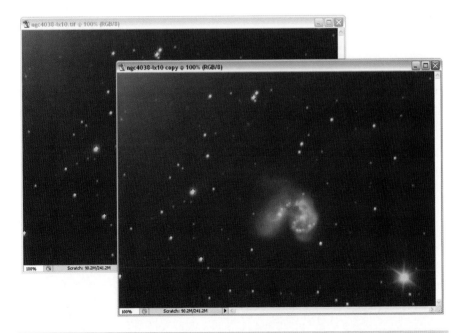

Fig. 6.10. Original and duplicate image of NGC 4038/39, showing some amplifier glow.

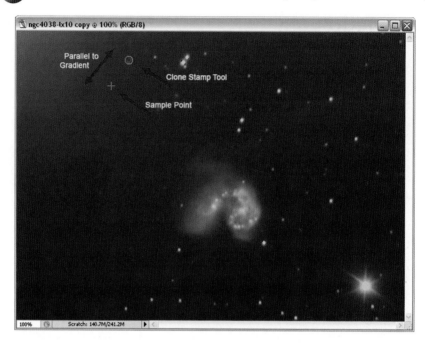

Fig. 6.11. Cloning tool getting rid of stars, etc., and showing the sample point in relation to the gradient.

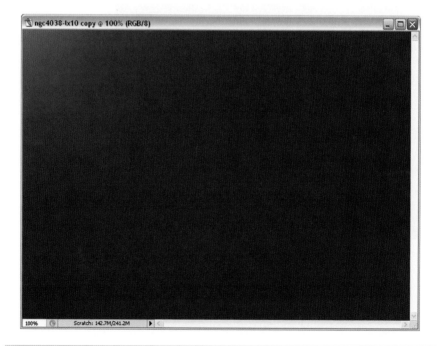

Fig. 6.12. All stars and objects removed.

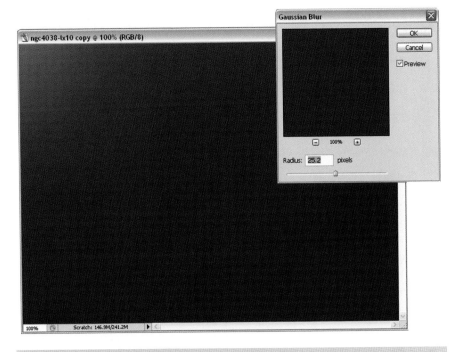

Fig. 6.13. Smoothing the gradient with a Gaussian Blur.

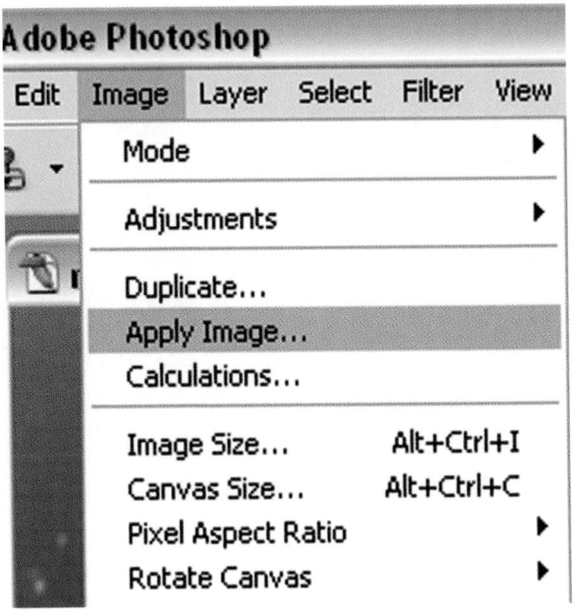

Fig. 6.14. Activate the Apply Image dialog box.

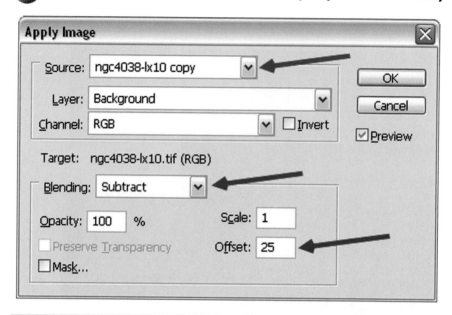

Fig. 6.15. Subtracting the gradient from the original.

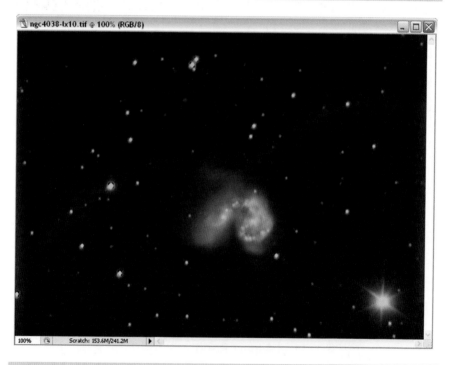

Fig. 6.16. Image now with a more consistent background.

(f) In the Apply Image dialog box …

1. Set the [Source:] to be the (blurred) copy image
2. Set the [Blending:] to Subtract
3. In the [Offset] box, try values from 1 to 30 (This value will vary from image to image (Fig. 6.15).)

Quick Tip: Left click and hold the mouse button over the word [Offset] and move the mouse left or right to change the value.

More recent versions of Photoshop have a really useful tool for fine-tuning certain applied actions. So, from the [Edit] menu, select [Fade Apply Image]. Moving the slider back and forth from 0 to 100, you see the before and after effect of the process, and it can then be set to whatever level is appropriate. Further adjustments can be made to the levels, histogram curves, and so on (Fig. 6.16).

It is worth saving the blurred mask for possible later use with similarly exposed images from the same optical system. An important constraint of the [Apply Image] function is that the two images be exactly of the same dimensions. When using the Duplicate or Select All and Copy functions, image sizes are replicated exactly.

6.4 Unsharp Masking

One of the most wonderful and widely used tools by amateur astronomers for processing images is the unsharp masking procedure. When used with care, this procedure can reveal an incredible amount of hidden structural detail in nebulae and galaxies.

Since many deep-sky objects typically have bright centers and faint extremities, revealing all the possible detail across the entire image requires a little trickery. Two classic examples are the magnificent Andromeda Galaxy (M31) and the great Orion Nebula (M42), where we commonly encounter glorious photographs showing an unavoidable, though heavily oversaturated central region in order to reveal the outermost, much fainter spiral arms or nebula.

A marvelous process for preserving detail in the bright regions of an image while at the same time showing all the faint details in the surrounding regions is to overlay a short exposure through a blurred layer mask onto a longer exposure. This approach is different from the classic "unsharp mask" filter commonly found in image manipulation programs that otherwise use a similar process (an algorithm) for sharpening detail. But these user-selectable filters do not recover information lost in over saturated regions of an image, and if used too aggressively can introduce or enhance unwanted dark contrast ring effects around stars.

How the following procedure works is by creating a fuzzy or blurry mask from a duplicate of the long exposure. The short and long exposures are blended together through this mask using the layer capabilities in Photoshop. By blurring the mask, it allows the short exposure to be combined to the long exposure without showing any abrupt changes in the brightness, contrast, or tone in the overlapping areas. These brightness, contrast, and tone values can be controlled during the process to produce a seamless merger of the two images (Fig. 6.17).

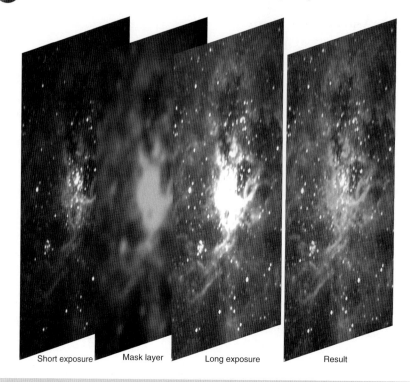

Short exposure   Mask layer   Long exposure   Result

Fig. 6.17. Blending short and long exposures through an "Unsharp" layer mask to preserve central detail.

There are two ways this process can be done when imaging targets that have very bright central regions.

1. A commonly used technique is to take two separate exposures of the target, one long exposure to record all the faint detail and one short to record the brightest areas without overexposure.
2. Working with a duplicate of the original image. This approach applies best to images taken with deep-sky capable video cameras, where the central detail is still evident in the original image but is lost when the levels are stretched during processing. Typically, parts of bright nebula, or the core of a globular cluster or galaxy, will become oversaturated after stretching the levels to reveal the faint outer regions of the object. By using the original image as the short exposure and stretching the levels of a duplicate to become the long exposure, the same principle is applied to recover the details lost in the stretching process.

The following procedure uses the second described method... working with a duplicate image.

6.4.1 Procedure

(a) Open the original image and make a duplicate (Fig. 6.18).

(b) Adjust the levels of the duplicate to bring out all the faint details; this will be the "long" exposure and the original will be the "short" exposure (Fig. 6.19).

(c) Make the original image active, press [Ctrl + a] to Select All, and press [Ctrl + c] to Copy. Make the duplicate image active and press [Ctrl + v] to Paste. Make the [Layers] palette visible (Fig. 6.20).

(d) In the [Layers] palette, make a mask by clicking the [Add Layer Mask] button (Fig. 6.21).

(e) In the [Layers] palette, click on the background layer, press [Ctrl + a] to Select All, and press [Ctrl + c] to Copy.

(f) *Now an important step.* In the [Layers] palette, while holding the [Alt] key, click the white mask box in Layer 1; the working image will go blank (Fig. 6.22).

(g) Now press [Ctrl + v] to paste the "long" exposure into the mask. This mask image will be seen in black and white even when working with color originals.

(h) From the [Filter] menu select [Blur – Gaussian Blur …].

Starting with the pixel radius value at zero, increase only just enough so that the fine details are blurred. Too little will not allow the foreground to blend smoothly, and too much will not hide the background enough. This image used a value of about 6. The value will vary for different images, so experiment. The result is our "unsharp mask."

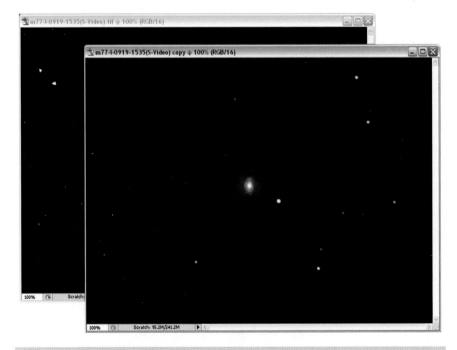

Fig. 6.18. Original and duplicate of the galaxy Messier 77.

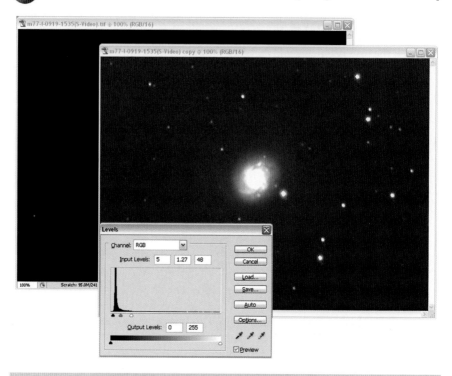

Fig. 6.19. Stretch the levels of the duplicate to become the "long" exposure.

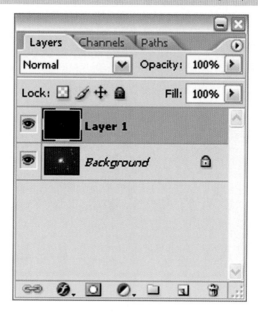

Fig. 6.20. The layers palette will show the new layer.

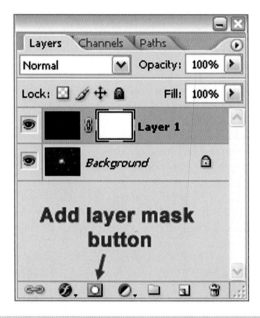

Fig. 6.21. Click the "Add Layer Mask" button to create a mask on the top layer.

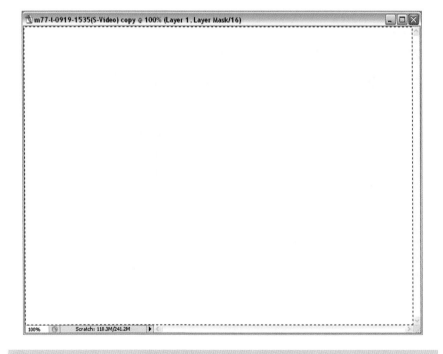

Fig. 6.22. The working image now shows as a blank window.

(i) To see the effects of the mask, from the [Window] menu select [Arrange – New Window], and another window will open showing the interaction of the three layers: "long" exposure, mask, and "short" exposure (Fig. 6.23).

Most of the time, this will be fairly close to the best values. If the over saturated part of the background image drops sharply to the surrounding areas, then some tweaking may be needed to blend this transition area in smoothly, if so, go onto Step (j). If not, proceed to Step (k).

(j) From the [Image] menu select [Adjustments - Brightness/Contrast] to fine tune the density and contrast of the mask. Start by reducing the contrast a bit and then adjust the brightness. Any changes to the mask will be immediately seen in the new window image (Fig. 6.24).

(k) When you are happy with the adjustments, from the [Layer] menu select [Flatten Image] and close the new window; the long-exposure image is now unsharp masked. From here further subtle adjustments can be made to levels, curves, saturation, and so on.

(l) Save your image as a TIFF file and denote it as being unsharp masked, that is, "m77-um.tif" (Fig. 6.25).

Figures 6.26 and 6.27 show two more applications of the unsharp mask procedure to maintain detail across an entire image. Give these procedures a try, and with a little practice, you will be amazed at what you might achieve!

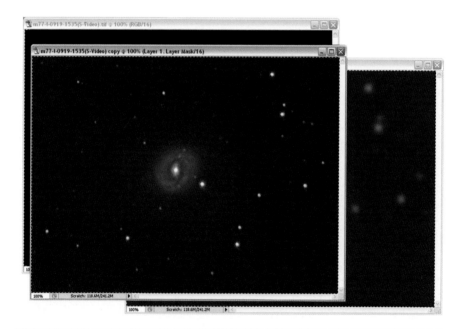

Fig. 6.23. Here we see the masking procedure at work.

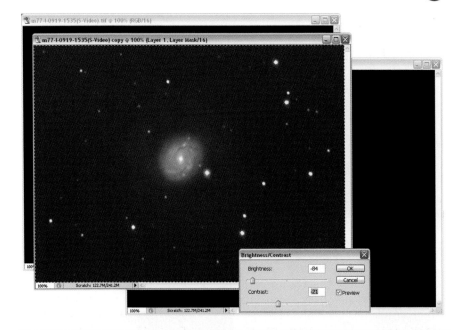

Fig. 6.24. Adjusting the mask contrast and brightness.

Fig. 6.25. The original, the stretched duplicate, and the final successful unsharp mask.

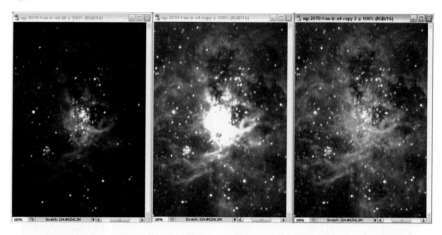

Fig. 6.26. Another example using the unsharp mask on the bright nebula NGC 2070.

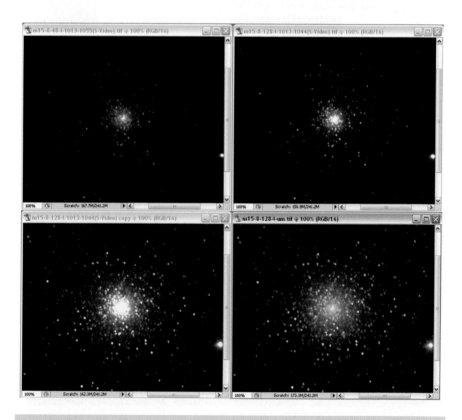

Fig. 6.27. This example, done in two stages, uses short, long, and stretched exposures with the unsharp mask routine on globular cluster Messier 15.

6.5 Making Color from Black and White Images

The magic of color! Yes, we all love a color picture and are constantly dazzled by the brilliance of pictures from the Hubble Space Telescope and even the highly talented amateurs across the globe using far more expensive cameras than our humble deep-sky video imagers. As we discussed in previous chapters, a monochrome camera is preferable for achieving maximum spatial resolution and for its overall greater sensitivity. Color images can be made by taking separate exposures through red, green, and blue (RGB) filters.

An alternative to standard dye-colored filters are dichroic filter sets comprised of cyan (for red filtering), magenta (for green filtering), and yellow (for blue filtering), which have higher overall light transmission characteristics and are often preferred by some amateurs for use in low light situations. After taking each separately filtered image they are then applied to the relevant color channel for each corresponding color component to produce a color 24-bit image.

So, if you have undertaken a separately filtered set of images using a monochrome video camera this is the procedure to bring the magic of color into one single image!

6.5.1 Procedure

(a) Open the RGB-filtered images (Fig. 6.28).
(b) Make the red image active and click on its [Channels] palette (Fig. 6.29).
(c) Make the green image active, press [Ctrl + a] to Select All, and press [Ctrl + c] to Copy. Make the red image active again, click on the green channel in its [Channels] palette, and press [Ctrl + v] to Paste (Fig. 6.30).
(d) Make the blue image active, press [Ctrl + a] to Select All, and press [Ctrl + c] to Copy. Make the red image active again, click on the blue channel in its [Channels] palette, and press [Ctrl + v] to Paste (Fig. 6.31).

You can close the green and blue filtered images now.

(e) It is advisable to save the image at this point as "*whatever object name*-rgb.tif" (Fig. 6.32).

Depending on how accurately you have taken the images, they may need to be aligned.

(f) To bring the three colors into alignment, while the blue channel is still highlighted in the [Channels] palette, click the "eye" box next to the RGB channel. This makes all the channels visible, but only the blue channel is active (Fig. 6.33). Zoom in to 200–300% to make it easier to see the alignment.
(g) Click the [Move Tool] icon and click back in the image. Use the cursor keys to line up the blue channel with the red.
(h) Click the green channel and repeat those steps to align the green channel with the other two.

Fig. 6.28. Red-, green-, and blue filtered images of NGC 5128.

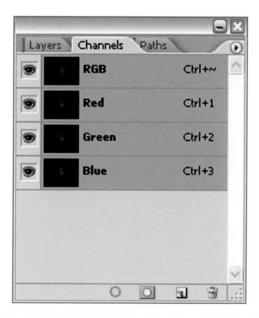

Fig. 6.29. Make the red image [Channels] palette visible.

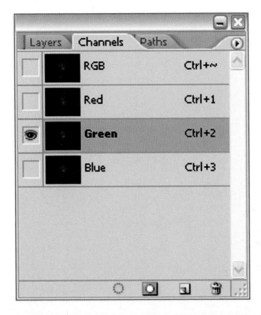

Fig. 6.30. Pasting the green image into the green channel of the red image.

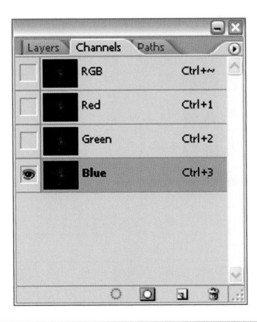

Fig. 6.31. Pasting the blue image into the blue channel of the red image.

Fig. 6.32. Save your new *RGB* image.

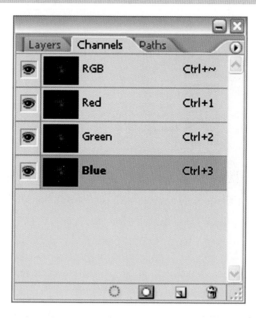

Fig. 6.33. All the "eyes" are on, but only the blue channel is active (*highlighted*).

Fig. 6.34. A correctly aligned *RGB* image.

You now have a well-aligned RGB image (Fig. 6.34).

Each channel can be selected and adjusted for brightness, contrast, etc., or click on the RGB channel and adjust all layers together. Save the image again when finished (Fig. 6.35).

Before using this image in the following procedures, make sure to click the RGB channel to ensure that all the color channels are active. (Same as Fig. 6.29).

6.6 Color Balance

The color produced in an image is open to personal preference and perception of the human eye. To achieve consistent color accuracy you have to rely on the numbers rather than the eye. To get absolute color you need to color calibrate your monitor using a reference star that is of the correct temperature, thus giving a pure white light.

For this procedure we offer this simple suggestion to achieve a fairly good and realistic color balance in your images. The method in Photoshop is by using the "Color Sample Tool." Exceptions to this procedure will yield varying results if different filters are utilized in place of a standard, calibrated RGB filter set.

ngc5128-rgb.tif @ 100% (RGB/16)

100%   Scratch: 204.6M/241.2M

Fig. 6.35. Adjusted levels *RGB* image.

6.6.1 Procedure

(a) Open your image, either single shot color or RGB composite (Fig. 6.36).
(b) Select the [Color Sampler Tool] by clicking and holding the [Eyedropper] tool until the submenu appears.
(c) Zoom in on the image for finer control. With the [Color Sampler Tool] set to a sample size of 3 × 3 or 5 × 5 average, click on the whitest star to set the white point. Make sure no other color influences the sample point. Sample points can be moved or deleted (right click on point, select delete).
(d) Next, click on the darkest part of the image that has no background nebula or vignetting influencing the sample point, and this will set the black point (Fig. 6.37).

Make sure the [Info] palette is visible. Note here that the RGB values differ (Fig. 6.38).

(e) From the [Image-Adjustments] menu, open the [Levels …] dialog box and select the red channel (Fig. 6.39).
(f) In the [Info] palette, a second number will appear next to #1R: 252/252. The objective here is to get the second number to read 255 by moving the right [input level slider] to the left on the histogram. Make sure you move the slider only enough so that the number just changes to 255. Any further and it will not change the number but will continue to alter the color. Then do the same for each of the green and blue color channels (Fig. 6.40).

Fig. 6.36. Original color image of Messier 17.

Fig. 6.37. Select within the whitest star and the darkest place on the image.

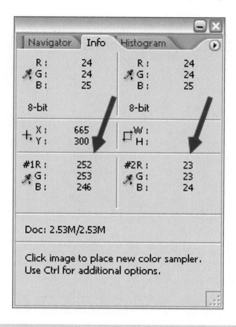

Fig. 6.38. Looking at the Info palette, note the *RGB* values of each sample point at #1 and #2.

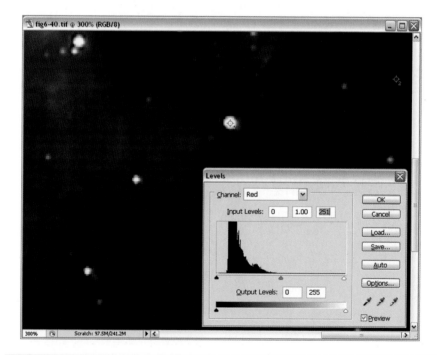

Fig. 6.39. Levels dialog box and red channel selected.

```
#1R :    252/ 255    #2R :    23/ 23
   G :    253/ 255       G :    23/ 23
   B :    246/ 255       B :    24/ 25
```

Fig. 6.40. All the *RGB* values now at 255.

```
#1R :    252/ 255    #2R :    23/ 30
   G :    253/ 255       G :    23/ 30
   B :    246/ 255       B :    24/ 30
```

Fig. 6.41. Dark point *RGB* values set to give a neutral background.

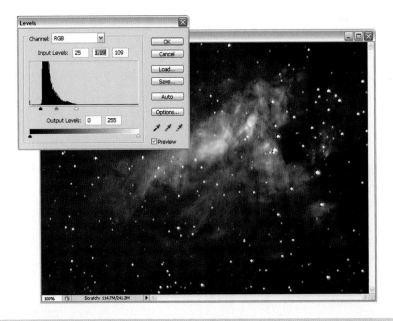

Fig. 6.42. Further adjustments can be made to the levels to bring out more information.

(g) Next move either the center [input level slider] or left [input level slider] (depending on the image and value at the second sample point) in each channel until the
#2RGB values all read the same. Depending on the image, aim for values of about
30–40. Setting this value to 0 will darken the image way too much, and you will lose
all the faint details. The night sky really is not totally black anyway (Fig. 6.41).

The image is now color balanced; while the [Levels…] dialog is still open, you can
make further adjustments, if required, to bring out more details, by selecting the
RGB channel. Save the image when finished (Fig. 6.42).

6.7 Adding a Luminance Channel to RGB

In Chap. 4, we discussed the incredible visual value that can be obtained by combining a non-color filtered image (a luminance exposure) with RGB exposures. Since filtered images are often subdued in overall brightness in a completed 24-bit RGB image, application of a luminance exposure has become common practice among many deep-sky astrophotographers. The image thereby becoming an LRGB image as opposed to an RGB then effectively reveals more extended detail across the entire subject with an overall brightening effect but in color! The practice is especially effective where faint objects are concerned.

In fact, seeing the results of a completed LRGB image is perhaps as rewarding as witnessing the sudden detail that leaped out of the initial stacked images during the middle tone grayscale stretching process covered earlier. So let us look at how we create an LRGB image in PhotoShop.

6.7.1 Procedure

(a) Open your RGB image and the luminance image (Fig. 6.43).
(b) Make the luminance image active and press [Ctrl + a] to Select All, press [Ctrl + c] to Copy. Make the RGB image active and press [Ctrl + v] to Paste. Make the [Layers] palette visible; it should now show the new layer.

Fig. 6.43. *RGB* and luminance (unfiltered) image of NGC 5128.

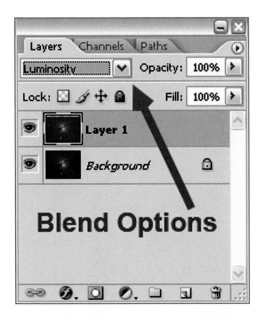

Fig. 6.44. Luminance layer Blend Option set to Luminosity.

(c) Align the new layer if necessary using the [Layers] palette blend option [Difference] and the [Move Tool], as previously described. Do not forget to click back on the image. Also, do not worry about the odd colors.

(d) Once in alignment, set the [Layers] palette blend option to [Luminosity] (Fig. 6.44).

We now have an LRGB image! You can simply click the eye next to layer 1 on and off to see the difference. The luminance layer now contributes all the fine detail and brightness information. In fact, the RGB layer can be blurred slightly to smooth out any color blotchiness, if needed (Fig. 6.45).

Again, the L or RGB layers can be selected and adjusted for brightness, contrast, and so on individually, or [Flatten Image] and work on all layers at once. Save the image and denote that it is an LRGB image, that is, "ngc5128-lrgb.tif" (Fig. 6.46).

6.8 Making a Simulated Green Channel

If you are imaging a subject in tri-color and time is of the essence, such as a fast approaching dawn, ugly dark clouds looming on the horizon, or you are simply feeling a bit lazy, then a simple short cut (a cheat) can be used to reduce the time needed to produce a full color image. By taking only red and blue exposures, these two images can then be applied to one another to produce a simulated green equivalent (an average) during post-processing.

Fig. 6.45. You now have an LRGB image.

Fig. 6.46. The final LRGB image with further levels adjustment.

Another application for this technique is for those who like to play around with Digitized Sky Survey (DSS) images. These are normally only available in blue and red exposures, so by making a simulated green, you can produce a nice color image for your desktop background.

6.8.1 Procedure

(a) Open the red and blue images. Make the blue image active, press [Ctrl + a] to Select All, and press [Ctrl + c] to Copy.
(b) Make the red image active and press [Ctrl + v] to Paste; the [Layers] palette should now show the new layer.
(c) Align the two layers if necessary using the [Layers] palette blend option [Difference], click the [Move Tool], click back on the image, and use the cursor keys to adjust.
(d) Once aligned, set the [Layers] palette blend option to [Normal] and the [Opacity] to 50% (Fig. 6.47).
(e) From the [Layer] menu, select [Flatten Image] and save as sim-green.tif.
(f) You can now create a color image by using the red, sim-green, and blue images as previously described in the "Making Color from Black and White images" procedure given earlier.

Although this will not be a true RGB image, it will give a good approximation of color. Adjusting each color channel can make the color balance look more natural. It is best to state that you have used this procedure when publishing the final image (Figs. 6.48 and 6.49).

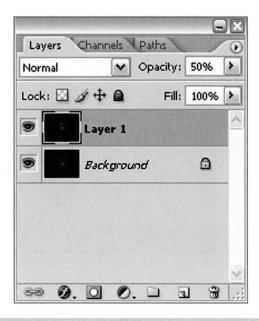

Fig. 6.47. Paste the blue image onto the red image and blend at 50% opacity.

Fig. 6.48. True red, green, and blue filters.

Fig. 6.49. Red and blue filters with simulated green channel.

6.9 Adjusting Color Shift in Single CCD Color Camera Images

While on the subject of color image creation and procedures for bringing individual color channels into correct registration, it is worthwhile mentioning shifted colors that are often produced by many single CCD color cameras. Aside from atmospheric and optical aspects that can add to color aberrations in an image, the effect we describe here is caused by spatial differences across the individually filtered pixels in the CCD array or the digitizing process.

Zoom in on a bright star in the image, and you will probably notice a red tinge at one end and a blue tinge at the other. The green channel is in between the two, while the red and blue channels are shifted by about one pixel on either side. By using the color channel registration method mentioned earlier, you can easily correct this slight color shift by simply nudging the red and blue channels by around one pixel towards the green channel, thus improving color registration and sharpness across the entire image.

6.10 Removing Artifacts

Most images will have some unwanted artifacts in them, be it hot pixels, dark spots after hot pixel removal, or dark rings around certain stars. These unwanted artifacts are pretty easy to get rid of using the [Spot Healing Brush Tool] and [Clone Stamp Tool] but require a bit of time and patience to finish an image off nicely.

Very Important Note: Great care should be exercised to ensure that you do not introduce false stars, bits of nebula, or other details that do not exist in reality. Equally important, do not delete any real information. Always strive to maintain the overall balanced look of an image. When applying any of these manual alterations to an image, make sure you do them equally to all parts, not just on smaller sections. These rules apply to all processing routines.

For this example you will notice that there is a consistent background around the dark spot, so the [Spot Healing Brush Tool] will work very well here. The Spot Healing Brush Tool automatically samples texture, lighting, transparency, and shading of the pixels surrounding the area you paint over. Zoom in to 200–300% for fine control. Select an appropriate tool size and paint over the dark spot with a little bit of overlap (Figs. 6.50–6.52).

If the artifact is near a star, then the Spot Healing Brush Tool will sample the star as well. In this situation, use the [Clone Stamp Tool] for greater control (Fig. 6.53).

In this example, make the sample points in line with the gradient of the background; sample from either side for an even smoother finish (Figs. 6.54 and 6.55).

An unfortunate drawback with video and certain other CCD-based cameras is the unwanted appearance of a dark ring and sometimes a bright ring pattern around the

Fig. 6.50. Dark spot after hot pixel removed in dark frame subtraction process.

Fig. 6.51. Using the Spot Healing Brush Tool, draw enough to cover the spot with a little bit of overlap.

Fig. 6.52. Artifact disappears with one click.

Fig. 6.53. Artifact near a star requires a sample point to be defined.

Fig. 6.54. Sample points (*crosses*) either side of the artifact.

Fig. 6.55. Artifact near a star removed with the Clone Stamp Tool.

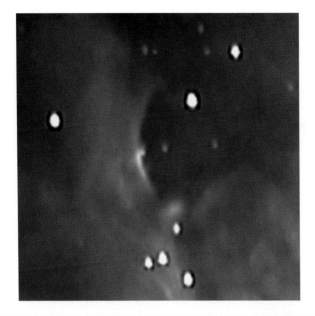

Fig. 6.56. Close-up of dark ring patterns around bright stars.

perimeters of bright stars (Fig. 6.56). These artifacts become evident especially when stretching brightness levels to their maximum.

If using the image for astrometry, photometry, or other scientific analysis, it should not be altered. If your goal is an aesthetically pleasing picture, then one or all of the following techniques can be utilized. Some images will respond differently to others in how you deal with these artifacts.

When sharpening an image, it may exaggerate these dark rings. The following reduction method, one of two given, works on the whole image.

6.10.1 Procedure 1

(a) Make a duplicate of the original image and perform sharpening on the duplicate.
(b) Make the original image active, press [Ctrl + a] to Select All, and press [Ctrl + c] to Copy. Make the duplicate image active and press [Ctrl + v] to Paste.
(c) In the [Layers] palette, set the blend mode to [Lighten]. From the [Layer] menu, select [Flatten Image]. Your image will have the benefits of the sharpening but with the dark rings highly reduced. This effect will be more or less apparent, depending on the amount of sharpening that has been applied (Fig. 6.57).

This next method works on individual stars that exhibit this effect to yield best overall aesthetic improvement, but it can be rather time consuming. But if you want a great portrait to show off to the world, then it is well worth the time.

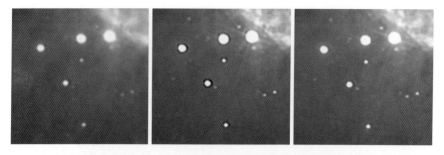

Fig. 6.57. Enlarged section of original image, sharpened duplicate, and combined image.

6.10.2 Procedure 2

(a) Zoom in on the image to 200–300% for finer control. Do not forget the space bar for moving the image around in the window. If a star with a ring has plenty of black sky around it, a fairly simple method can remove most of it (Fig. 6.58).

(b) Using the [Elliptical Marquee Tool], draw a circle around the star, dark ring, and bright ring.

Quick Tip: Set the [Style:] of the [Elliptical Marquee Tool] to [Fixed Aspect Ratio], and it will always draw a perfect circle (Fig. 6.59).

(c) Press [Ctrl + c] then [Ctrl + v] and set the [Layers] palette blend mode to [Color Burn] (Fig. 6.60).

(d) Adjustments can be made with the [Opacity] of the layer.

(e) When you are happy with how it looks, from the [Layer] menu select [Flatten Image] (Fig. 6.61).

Ultimately, to achieve an aesthetically pleasing result, fine work with the [Clone Stamp Tool] is required. By "nibbling," or taking multiple sample points all around the dark ring, you can produce a pleasing finish. Make sure you smooth out the bright ring as well for a smooth blending of the background. This technique does need a bit of practice but in time becomes almost second nature (Figs. 6.62 and 6.63).

After the nibbling process is complete, you can yield an even finer finish using the [Radial Blur] filter. This is found in the [Filter] menu under [Blur …] and will smooth any unwanted jagged-looking edges. In the [Radial Blur] filter dialog box, try "amount" around 70% for starters and use "Blur Method" Spin. Make sure the selection around the star is circular (Fig. 6.64).

On some very bright stars the nibbling "in" technique may not work quite as well, so try nibbling "out" instead, to give the star a fuller appearance. Work with the [Clone Stamp Tool] as before, hold the [Alt] key, and select the center of the bright star as the sample point. Untick the [Aligned] checkbox in the [Options] bar near the top of screen. Fill in the dark ring around the star. Again the [Radial Blur] filter can be used to give a smoother finish to the edge of the star (Fig. 6.65).

For the adventurous, you can play around with the multitude of functions found in Photoshop and create your own solution to correct these artifacts, for example, with the use of multiple layer masks.

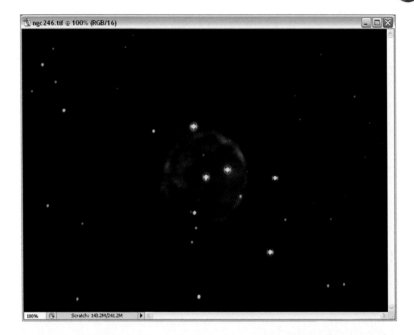

Fig. 6.58. Original image of NGC 246 with some dark and bright rings around the brighter stars.

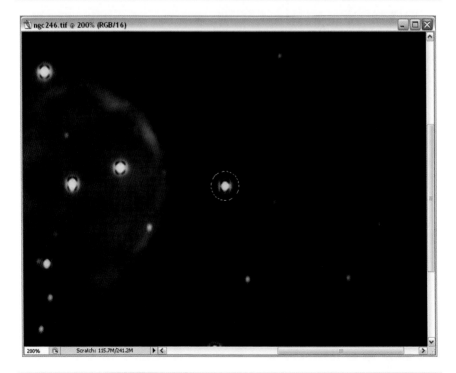

Fig. 6.59. Selecting around the dark ring and bright ring.

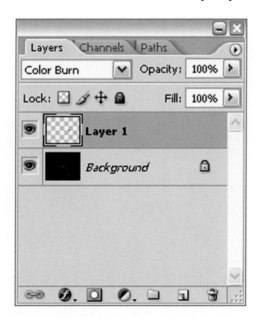

Fig. 6.60. Set the layer blending to Color Burn.

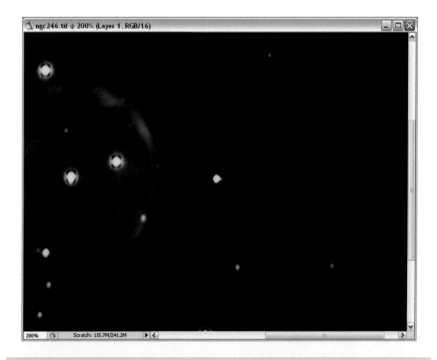

Fig. 6.61. Most of the ring is removed.

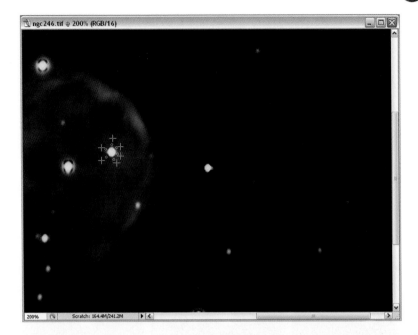

Fig. 6.62. Nibbling around a star to remove dark and light rings; crosses show the sample points.

Fig. 6.63. All the dark rings removed for an aesthetically pleasing finish.

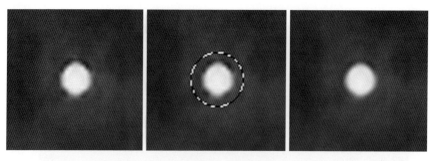

Fig. 6.64. Fine finish with the Radial Blur filter.

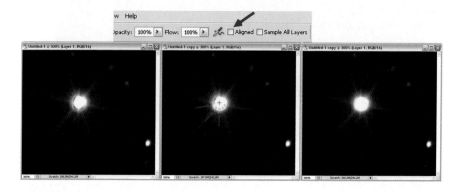

Fig. 6.65. Reverse nibbling a very bright star.

Like many other innovative amateurs with software programming skills that led them to creating useful freeware tools for astronomy, a neat little French-authored program called "Loreal" does a great job most of the time (though with a little experimentation) at removing these dark rings around stars. At the time of this writing, though, the menus were still in French.

6.11 Blending MultiField Images

We discussed earlier in Chap. 4 how some amateurs like to produce larger scale images to maintain resolution by imaging two or more sections of the sky surrounding a certain object of large angular size, that is, M31, M42, NGC253, and so on. Several image manipulation programs employ automatic merging features to stitch images together. Well, the ever popular Photoshop has one also, which can be found under the [File – Automate – Photomerge …] menu.

For greatest ease of control over the final result, we suggest the following manual procedure.

6.11.1 Procedure

For this exercise we will blend two separate fields to make one wide image.

(a) Open both images (Fig. 6.66).
(b) Make the left image active and open the [Image - Canvas Size …] dialog box (Fig. 6.67).

Fig. 6.66. Two separate fields showing each half of the galaxy Messier 31 with an overlap of about 15%. Imaged through a 300-mm lens with a GSTAR-EX camera.

Image	Layer	Select	Filter	View
Mode				▶
Adjustments				▶
Duplicate…				
Apply Image…				
Calculations…				
Image Size…		Alt+Ctrl+I		
Canvas Size…		Alt+Ctrl+C		
Pixel Aspect Ratio				▶
Rotate Canvas				▶

Fig. 6.67. Open the Canvas Size dialog box.

(c) Click the middle-left anchor point and make the width 200%. This will extend the image background (Fig. 6.68).

(d) Make the right image active and, using the [Rectangular Marquee] tool, draw a selection box over the image, leaving a small gap on the left edge (at least 25 pixels for this example).

Quick Tip: An easy way to do this is by pressing [Ctrl + a] and moving the selection box to the right with the cursor keys. This ensures that the selection is all the way to the top and bottom of the image and, therefore, only the left edge will be "Feathered" next.

(e) Open the [Select – Feather …] dialog box and set the [Radius] to 20 pixels (different images may need other values). This will give the overlapping part a soft edge that will blend in smoothly (Fig. 6.69).

(f) Press [Ctrl + c] to Copy this selection, make the left image active, and press [Ctrl + v] to Paste.

(g) Set the [Layers] palette blend mode to [Difference] and align the overlapping parts of the two images. If necessary, use the transform controls to rotate or stretch the layer. Press [Enter] to finalize the transform (Fig. 6.70).

(h) Once aligned, set the [Layers] palette blend mode back to [Normal]. You should see a smooth merger of the two images.

(i) Open the [Image – Adjustments - Brightness/Contrast …] dialog box and adjust until the two layers have the same background tonal qualities.

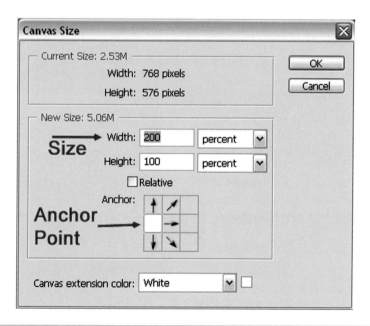

Fig. 6.68. Choose the Anchor point and set width to 200%.

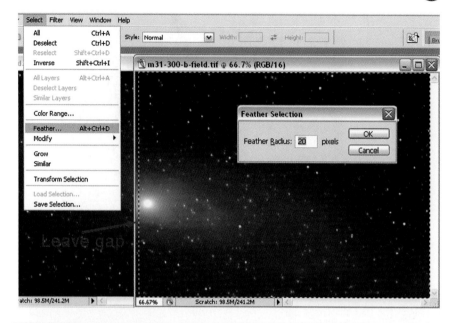

Fig. 6.69. Selection with Feathering set to 20 pixels.

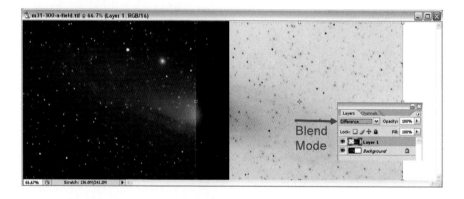

Fig. 6.70. Align the two overlapping parts.

(j) When you are satisfied that the merger is seamless, from the [Layer] menu select [Flatten Image] and crop the excess from the edge of the image. Save as a new image, that is, "m31-stitched.tif" (Fig. 6.71).

This final image has unsharp masking and further adjustments to the levels to bring out all the information (Fig. 6.72).

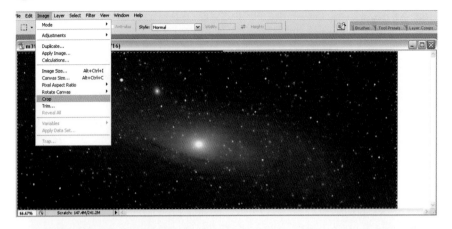

Fig. 6.71. Crop the excess from the image.

Fig. 6.72. Final adjustments to bring out all the subtle details.

6.12 Printed Video Images

It is great viewing your results on a computer monitor, where the inherent low-resolution performance makes most any picture look great, but a nice printed version of your hard-earned portrait framed on the wall at home or submitted for printing in an amateur astronomy magazine picture gallery is also very rewarding.

Making nice hard copy prints of low-resolution video images is not as demanding as the strict printer industry requirements of years ago. Yes, there is a practical rule of thumb giving 300 dpi (printed dots per inch) as the minimum resolution required for the human eye to perceive a photorealistic picture. Depending on your video camera, capture device, and computer, the typical per inch resolution of most images produced at the computer is 72–96 dpi. Therefore, at 72 dpi as viewed on your computer monitor of the same resolution, your image must be at least 4.16 inches across to be reproduced at a print quality (photorealistic) size of 1 inch. But, with new and improved printing processing and even simple bicubic scale up resampling to a certain point, most such images can be reproduced very nicely at larger scales, especially where modern consumer-based inkjet printers are concerned.

In fact, we have found that some inkjet and laser printers using top quality glossy papers can produce very nice results, defying these rules up to 8 × 10 inch (A4) prints with fantastic results. Sure, if you want to go over the print with a magnifying glass seeking out defects, go ahead, but in reality, most people will simply stand back and admire the picture with little or no awareness of the boring "must adhere to" details.

In some cases, it may be necessary to brighten the image on screen so as not to lose valuable detail when printed. This will depend on how well your printer is calibrated in accordance to what you are seeing on the computer monitor.

A handy tip where inkjet printing is concerned is to resize/resample your image. By making the image size 200% larger with Bicubic Resampling, the "blocky" appearance of stars and fine detail will be smoothed out. Then, when you print, use the [Print with Preview ...] dialog box to scale the image back to fit the paper size.

In Photoshop this command is found under the [Image – Image size ...] menu. Set the [Resample Image] option to [Bicubic Smoother], make sure [Constrain Proportions] is ticked, and make the width 200% (Fig. 6.73).

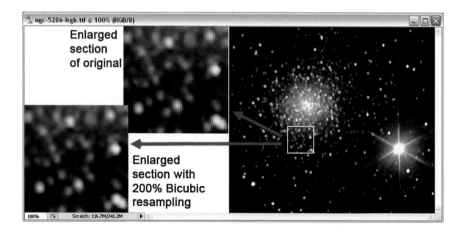

Fig. 6.73. NGC 5286 is seen here as an example of bicubic resampling. The enlarged sections show the difference with and without resampling.

6.13 Publishing Images to a Website

When publishing images to a Website you should be aware that the originals will be quite large files, and not everyone has high-speed internet. Reducing the size to fit most browser windows and compressing the image will keep files manageable for downloading.

Image sizes of between 500 and 800 pixels wide will fit nicely on a Webpage without the need for scrolling to see the whole image. Use Joint Photographic Experts Group (JPEG) compression to keep the file size to around 100–150 kb or less. You can always have a link to the full-size image if people are interested.

An excellent tool in Photoshop is the [File – Save for Web …] command. This opens a dialog box with adjustments for file type, image quality, and image size. The compression algorithm is very good at maintaining detail in JPEG images while also retaining a small file size.

Also, do not flood your individual Webpages with too many images, as people become impatient while waiting for all the pictures to load. If you have lots of images to show off, then divide them into separate categories and use thumbnail images as a sampler for the full-size image behind it. When we say thumbnail images, we mean a "true" separate, down-scaled resample of the master image for quick previewing.

A common lazy practice sometimes encountered is to simply use only the full-size master image and reduce it down to thumbnail size using the corner click and size scaling tool of the HTML Webpage editor. Although it may look small in the browser, it is still the large master image behind it being read in to the thumbnails preview page and, therefore, takes just as long to load as it takes to view the full-sized version anyhow. Trying to view a thumbnail page poorly designed in this way can be very frustrating, especially if the images posing as previews are 500 K and larger.

Thumbnails can be compressed quite significantly while still looking very appealing overall. So it is worthwhile making the effort. Visitors to your site will be grateful.

CHAPTER SEVEN

Other Video Applications

7.1 Lunar and Minor Planet Occultations

The sensitivity of modern surveillance-type video cameras is ideal for occultation work, where the fast frame rates can yield subsecond accuracy to these observations. Live images through moderate-sized telescopes can easily reveal stars down to 12th–14th magnitude, and coupled to timing devices such as KIWI OSD, which utilize global positioning system (GPS) signals, accurate asteroid and lunar occultation timings are made pretty simple (Figs. 7.1 and 7.2).

When an asteroid passes in front of a star we see a stellar eclipse or occultation. Knowing the precise time of when the shadow passed over a given location on earth will produce a very accurate position for the asteroid. If several observers time the same event from different locations (cords), a profile of the asteroid can also be deduced. Tiny asteroid moons have also been discovered by using these techniques. Using a PAL video camera recording at 25 fps and reviewing the individual odd and even interlaced frames, you can resolve the time of an event to ±20 ms (the length of an exposure). With NTSC format this is 16.6 ms. With further Fresnel diffraction analysis of the measured points of brightness in a graph of the event, even greater accuracy can be achieved (Fig. 7.3).

S. Massey and S. Quirk, *Deep-Sky Video Astronomy*,
DOI: 10.1007/978-0-387-87612-2_7, © Springer Science + Business Media, LLC 2009

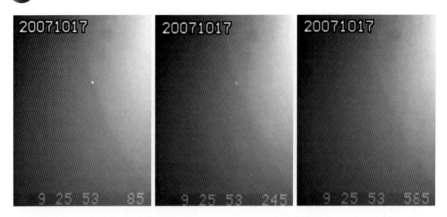

Fig. 7.1. On October 17, 2007, the moon occulted this close double star ZC2601 (7 Sgr, magnitude 6.7), and you can see, very faintly, a residual image for a number of frames after the main disappearance (Still 2). The separation of the two stars is only 0.2 of an arc second. The numerals at the bottom are the *GPS* time stamp by KIWI OSD. Courtesy of Steve Kerr.

Lunar Occultation (R) of chi Aquarii   by Dave Gault

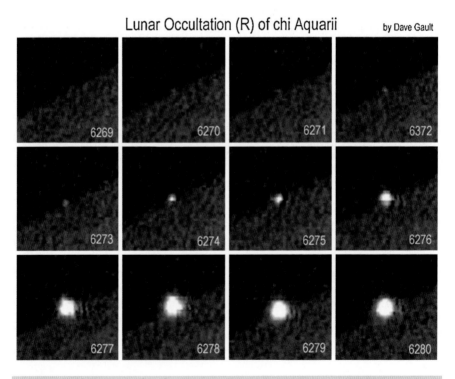

Fig. 7.2. This sequence shows a star, Chi Aquarii, reappearing from the edge of the moon. Numbers in each frame show the timing in milliseconds. Courtesy of Dave Gault.

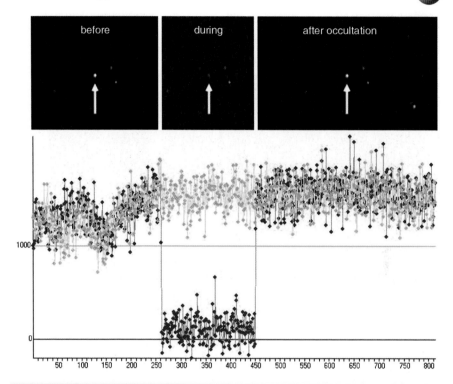

Fig. 7.3. On June 6, 2008, this asteroid occultation was observed from Mudgee, NSW Australia. Asteroid 733 Irmintraud occulted star UCAC2 12089544 in the southern constellation of Ara. The shadow event started at UT 12:33:51.74 and ended 7.56 s later at UT 12:33:59.3. The *blue points* on the graph show the brightness of the asteroid and star, then only the asteroid during the occultation, then the asteroid and star again. The *yellow points* show the brightness of a nearby comparison star.

7.2 Astrometry

Supernova, comet, and asteroid astrometry are other areas where deep-sky video works beautifully, and it is not just a fun project but one of real scientific value. Many amateurs worldwide contribute by following up on newly discovered asteroids and comets, allowing their orbits to be determined quickly. The small field-of-view images produced with CCTV video cameras, in particular, yield quite accurate positional data when used with programs, such as Astrometrica (Figs. 7.4–7.6)

Asteroid position work also opens up thousands of targets to moderate-size telescopes. One possible drawback of small video chips is that the limited field of view may have too few astrometric reference stars. This can reduce the accuracy of

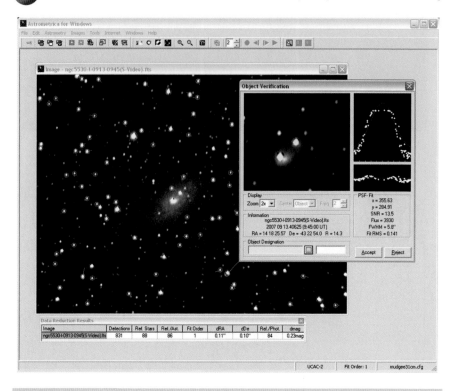

Fig. 7.4. Confirmation image of SN 2007it in galaxy NGC 5530, discovered by Bob Evans in August 2007. Position accurate to 0.01 of an arc second was submitted with the discovery notification.

the result, but some star catalogs (such as USNO-B1.0) go very faint, and although not the most accurate catalogs, they should almost always allow an astrometric solution to any field.

Astrometry is a useful means of determining the focal length, f/ratio, CCD orientation, and image scale. Double stars can be followed in their orbits using long focal lengths doing astrometry systematically over many years.

7.3 Supernova Searching

Using a live image to compare galaxies with your previously recorded results or a DSS image is an easy way to search for extragalactic SN. With moderate-size telescopes and maximum accumulation mode, the live video image can reveal stars down to around 15th–17th magnitude, opening up greater opportunities to find these elusive explosions. Alternatively, a quick stacked image of only 20 to 25 maximum

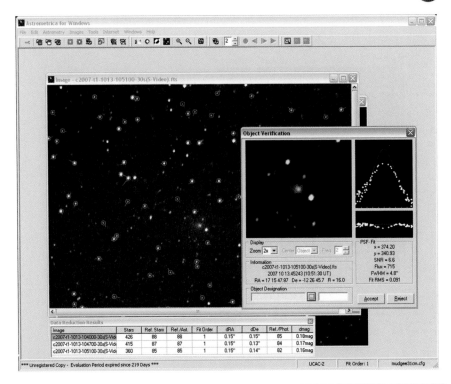

Fig. 7.5. 14th magnitude Comet McNaught 2007T1 gave positions accurate to 0.1 of an arc second with only 30 seconds recordings. 2007-Oct-13.44809 (10:45:15 UT) RA = 17h 15m 47.98s Dec = –12d 26m 33.8s Rmag = 15.9

accumulation frames can go even fainter. Even searching under bright moonlit skies is possible where conventional visual searching is not as effective (Fig. 7.7).

7.4 Meteor Observations

When coupled with a PC running meteor recognition software, such as Metrec and a wide-angle lens, video cameras are great for recording meteors down to about the 3rd magnitude in real time (Figs. 7.8 and 7.9).

7.5 Video as a Guiding Tool

If you want to do wide-field long exposure photography, be it with a specialized "cooled" camera, film, or DSLR, then accurate tracking is essential in order to achieve nice round or pinpoint stars. With a well polar-aligned mount, video has

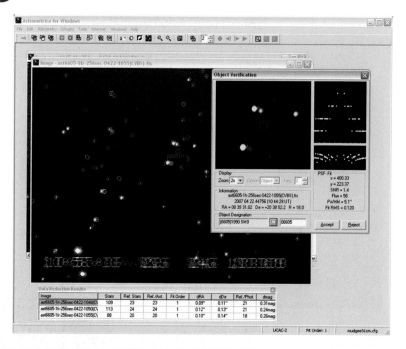

Fig. 7.6. 18th magnitude Asteroid 6605 yields a position on 2007 04 22.44756 of RA = 08h 39m 31.02s Dec = +20d 38m 52.2s (Though hardly visible in this image, it was detectable by the software).

Fig. 7.7. 12th magnitude SN2006dd only several arc seconds north of the tenth magnitude nucleus of NGC 1316 (these images have South at the top). It was easily visible on the monitor even with the full moon still quite high in the sky. Stars fainter than 16th magnitude are recorded even under these conditions. *Left*: 200-frame stack with no moon. *Right*: 50-frame stack during full moon.

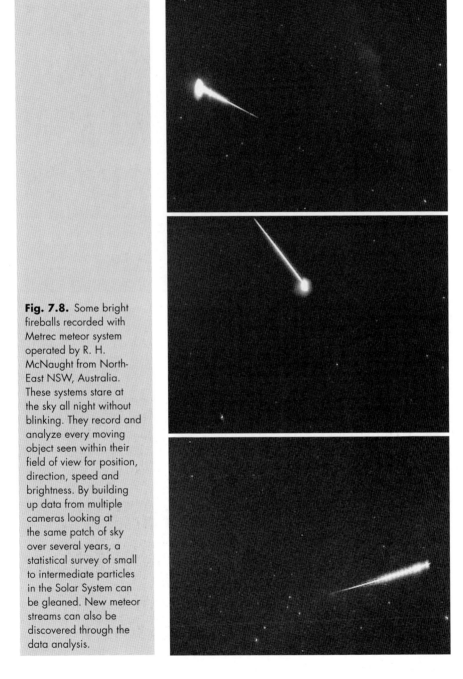

Fig. 7.8. Some bright fireballs recorded with Metrec meteor system operated by R. H. McNaught from North-East NSW, Australia. These systems stare at the sky all night without blinking. They record and analyze every moving object seen within their field of view for position, direction, speed and brightness. By building up data from multiple cameras looking at the same patch of sky over several years, a statistical survey of small to intermediate particles in the Solar System can be gleaned. New meteor streams can also be discovered through the data analysis.

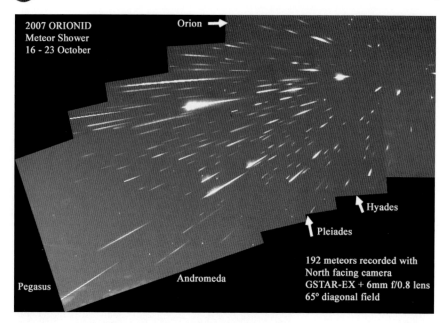

Fig. 7.9. Combined image of 192 Orionids recorded in 2007 with the same system as Fig. 7.8.

proven to be a very effective tool for both automatic and manual telescope guiding. With its low-light sensitivity and fast accumulated frame output, many more faint guide stars can be utilized for tracking purposes than are normally possible using visual guiding with a crosshair reticle ocular.

7.6 Manual Guiding

If you are using an accurately aligned piggyback scope for guiding then your deep-sky capable video camera is simply placed in the focuser instead of the otherwise redundant illuminated crosshair eyepiece. If using a separate TV monitor as the guide star viewing medium, you can simply place a crosshair over the CRT display using a non-permanent felt pen or thin strips of intersecting paper tape. The guide star (if bright enough) should still be visible through the paper tape. If using a computer with a capture device for guiding, moveable crosshair generating software can be used for gauging star drift and making corrections with the hand controller (Fig. 7.10).

Fig. 7.10. GSTAR-Capture software with crosshairs on screen.

7.7 Automatic Guiding

Today, there are several autoguiding options that measure the drift of the guide star image and then send commands to computerized mounts to make the necessary tracking adjustments. Some are available as software programs such as GuideDog, which require a video capture device to monitor the X-Y pixel position of a selected guide star, then sending the necessary commands via a cable from the computer's communication port back to the mount. But, depending on your computer's capacity to deal with multiple processes for handling both the video capture and the camera doing the long exposure, there can be conflicts within a purely software-managed environment. If this is the case, then there are other practical options, such as a standalone TV guider like SBIG's STV system and others.

Martin Myslivec from the Czech Republic developed a very neat and compact TV guider system that works exceptionally well with compatible and popular commercial mounts, including Meade, Celestron, Losmandy, Vixen and the popular HEQ5 and EQ6 by Synta. The model we tested came with a 9-cm LCD screen and supporting video and coms interfacing cables. Powered by 12VDC, it includes several touch button controls and input/output ports for computer and camera interface.

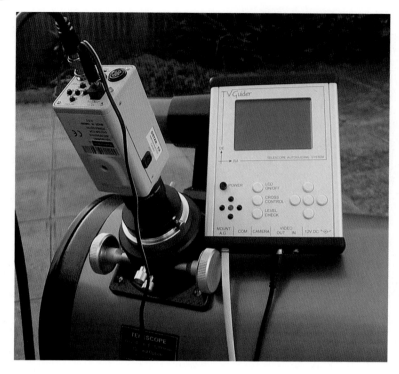

Fig. 7.11. The nicely constructed and affordable TV Guider system developed by Martin Myslivec of the Czech Republic, interfaced to the autoguider port of an EQ6 equatorial mount.

Just like a software-based guiding program, its firmware can be upgraded from downloadable Web updates using its com port connector. One of the nicer features of most stand-alone systems like these is the video input and output loop, which allows cameras with only a single video-out connector to be used for guiding and imaging simultaneously (Figs. 7.11 and 7.12).

7.8 Video Finderscope

Yet, another very practical application of the frame accumulation video camera is locating objects long before we can see them visually through an optical finderscope shortly after sunset. You will be amazed at how well these cameras work in such situations. Fitted with a C/CS mount lens and mounted piggyback on the main optical tube, you can carry out GOTO star alignments or simply target bright objects in the early dusk hours more efficiently than waiting until the sky has darkened enough for visual detection. For Mercury observers this can be very useful for detecting and viewing the planet while it is still at higher elevations above the horizon. This also applies to bright planetary nebulae such as the Ghost of Jupiter (Fig. 7.13).

Fig. 7.12. DSLR image of NGC4945 by Allan Gould, using a Watec video camera with 70 mm guide scope and TV Guider system connected to a HEQ5 equatorial mount.

Fig. 7.13. A video finderscope setup by Albert Van Donkelaar.

7.9 Video Polar Alignment

Polar alignment of a telescope mount can be done roughly or accurately. Some amateurs use a polar axis scope for quick set up or an illuminated crosshair reticle for refining alignment based on visible drift in each axis. In the case of the latter, especially where a polar scope may be useless with obstructed views to the north or south celestial pole, your deep-sky video camera can also be used to facilitate accurate polar alignment before you start imaging with it. And this makes the task far more relaxing than squinting hunched over into a reticle eyepiece.

Using either taped, felt pen drawn, or software-generated crosshair reticles on the video monitor, the eye-relief is unsurpassed, and the large image scale of small CCD chips yield ample magnification, perfectly suitable for quick detection of any directional drift. By orienting your video camera so that the horizontal crosshair matches right ascension adjustments and the vertical crosshair matches declination movement you will quickly be able to fine tune your mount to counter the drifting motion seen in each axis.

7.10 Collimating with Video

Using the video camera in place of your eye while collimating your telescope, will show the alignment of the optics on a monitor. The monitor can be placed in a convenient position and viewed easily while adjusting the collimating screws at the rear of the telescope. The video also gives higher magnification than just the eye alone, allowing for finer adjustment (Fig. 7.14).

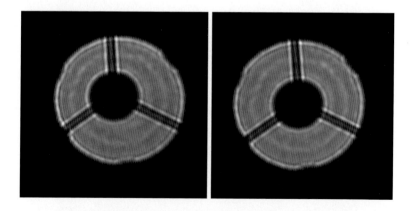

Fig. 7.14. Using video to correct the collimation of telescope optics. *Left* – out of collimation; *right* – optics properly aligned.

7.11 Detecting Faint Planetary Moons

Although solar system objects are not technically deep-sky related in topical terms, we do consider faint targets such as many small asteroids and comets to fall into this category. Likewise, there are those visually extreme, faint moons that orbit the planets Mars, Saturn, Uranus, and Neptune, which certainly deserve mention, given the capabilities of frame accumulation video cameras today. Aside from the four bright Galilean moons, perhaps a few of the brightest within Jupiter's armada of small, faint moons may be possible to image under ideal circumstances.

Seeing the moons of Uranus or Neptune at the eyepiece of small- to medium-sized telescopes, for example, is virtually impossible in most circumstances. But turn on the frame accumulation mode of your deep-sky capable video camera only a few steps, and there they are on the video monitor!

The tiny moons of Mars, Phobos and Deimos are renowned optical challenges, but when separated enough from the mother planet, they can be detected with a frame accumulation video camera. Even those newly discovered planets beyond the orbit of Pluto can be revealed (Figs. 7.15 and 7.16).

Fig. 7.15. *Top:* Mars overexposed in center with Phobos and Deimos on either side. *Bottom left:* Neptune and Triton imaged over two nights. *Bottom right:* Tri-color composite showing disc of Uranus and deeper exposure to reveal four of its moons.

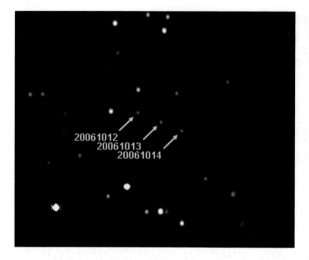

Fig. 7.16. Distant objects such as the dwarf planet Eris imaged over three consecutive nights. Each night it was imaged through a 31.5 cm f/4.5 Newtonian reflector and GSTAR–EX video camera. Each position is a stack of 400 × 2.56 second viedo frames. Eris was magnitude 18.7 at the time.

7.12 Near Infrared: Imaging the Unseen

Other deep-sky targets that are open to us due to the unfiltered infrared (IR) sensitivity of the monochrome CCD are those that appear bright in the IR part of the spectrum. Apart from being high IR emitters, many of these objects show little brightness at visual wavelengths (Fig. 7.17).

And, although also not technically falling in to the deep-sky focus of this book, we feel it worthwhile to mention the IR capability of these cameras in revealing one particular planetary aspect normally invisible to the human eye - Venus.

Venus is one unique target generally considered boring at visual wavelengths, due to its thick canopy of highly reflective cloud cover surrounding the entire planet. More interesting are efforts to image it using an ultraviolet filter to reveal various subtle patterns within the upper cloud deck.

However, since the planet is so incredibly hot at its surface, some of the heat from its surface is re-radiated through its clouds and back into space. Discovered by ground-based astronomers in the 1980s are a handful of narrow IR windows at wavelengths 1.0 μm, 1.7 μm, and 2.3 μm, which allow us to peer through the planet's dense cloud layers to reveal the structure of deeper cloud layers and regions of slightly cooler

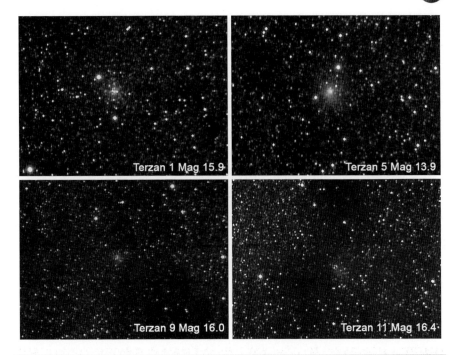

Fig. 7.17. Very faint and highly reddened globular clusters, such as these members of the Terzan group in Sagittarius and Scorpius, are fair game for the *IR* sensitive video camera.

topography. Lowlands appear brighter, while upland regions (being slightly cooler) appear darker.

Since many monochrome CCD cameras have a spectral sensitivity sloping off to around 1.1 μm, use of a 1-μm filter will indeed reveal the planet's night side when its sunlit portion is passing through the thin crescent phases. Because most CCDs are less sensitive at this extreme end of their spectral response, the image produced is very faint, and, therefore, setting the camera's accumulation mode from, say, X8 to X24 will then reveal the entire disk of the planet similar to the effect of earthshine seen during crescent moon phases (Figs. 7.18 and 7.19).

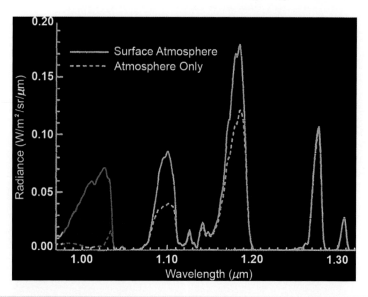

Fig. 7.18. Graph of infrared windows used to image Venus. The red boxed area shows the 1-micron window which certain video cameras are capable of recording.

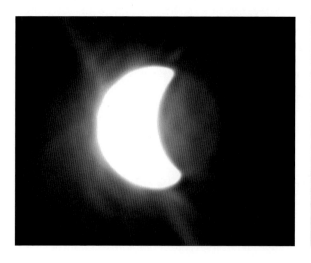

Fig. 7.19. The night side of Venus revealed with a 10-inch Newtonian and Mintron frame accumulation camera fitted with a professional grade 1-micron filter on the 4th July 2004. Image by S. Massey

CHAPTER EIGHT

The Gallery

The following images should inspire you to explore the enormous potential of low-light sensitive frame-accumulating video cameras. The results are divided into two broad groups: Suburban Light-Polluted Skies and Dark Country Skies.

All images were taken with the basic GSTAR-EX camera to show what can be achieved with minimal frame accumulation rates in an affordable monochrome camera. The images also serve to illustrate just how well short-integration exposures can produce great deep-sky images with some later processing in Adobe PhotoShop.

The city suburban images should hopefully serve to show just how accessible many deep-sky targets are with these cameras. All frames for the following images were captured at 128X sense-up (unless otherwise noted).

8.1 Suburban Light-Polluted Skies

(Figs. 8.1–8.20)

S. Massey and S. Quirk, *Deep-Sky Video Astronomy*,
DOI: 10.1007/978-0-387-87612-2_8, © Springer Science + Business Media, LLC 2009

Fig. 8.1. Messier 20 in Sagittarius – 25 cm f/4.7 Newtonian, Vixen 0.6X focal reducer, 500 × 2.56s (L), 250 × 2.56s (of each RGB). S. Massey.

Fig. 8.2. NGC 253 in Sculptor – 25 cm f/4.7 Newtonian, Vixen 0.6X focal reducer, 500 × 2.56s (L), 200 × 2.56 (of each RGB). S. Massey.

Fig. 8.3. Messier 27 in Vulpecula – 25 cm f/4.7 Newtonian, 250 × 2.56s (of each RGB).
S. Massey.

Fig. 8.4. Messier 17 in Sagittarius – Meade SN8, 480 × 2.56s (of each RGB). Courtesy
of Mark Garrett.

Fig. 8.5. NGC 5139 in Centaurus – 80 mm *f*/7 ED refractor, 200 × 2.56s (of each RGB). S. Massey.

Fig. 8.6. Messier 8 in Sagittarius – Meade LXD75 SN8 at *f*/4, 101 × 2.56s (L). Courtesy of Mark Garrett.

Fig. 8.7. NGC 4945 in Centaurus – 25 cm f/4.7 Newtonian, 150 × 2.56s (L). S. Massey.

Fig. 8.8. Messier 42/43 in Orion – ProStar 127 mm f/7.5 APO refractor with Vixen 0.6 focal reducer, 450 × 2.56s (L), 150 × 2.56s (of each RGB). S. Massey.

Fig. 8.9. Barnard 33 in Orion – 25 cm f/4.7, 150 × 2.56s (L). The Horse-Head Nebula is a very difficult target especially from light-polluted areas like Sydney. S. Massey.

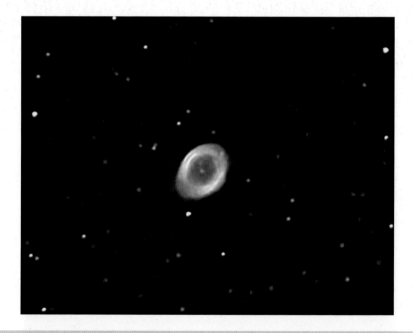

Fig. 8.10. Messier 57 in Lyra – 25 cm f/4.7 Newtonian, 300 × 2.56s (L), 300 × 2.56s (of each RGB). S. Massey.

Fig. 8.11. NGC 4565 in Coma Berenices – 25 cm f/5, 550 × 2.56s (L), last-quarter moon in the sky. S. Massey.

Fig. 8.12. Messier 16 in Serpens – 25 cm f/5 Newtonian, 700 × 2.56s frames, Red + LPR + IR block filters (heavy light pollution). Courtesy of Darrin Nitschke.

Fig. 8.13. NGC 3372 in Carina – 25 cm f/4.7 Newtonian, 300 × 2.56s (of each RGB). S. Massey.

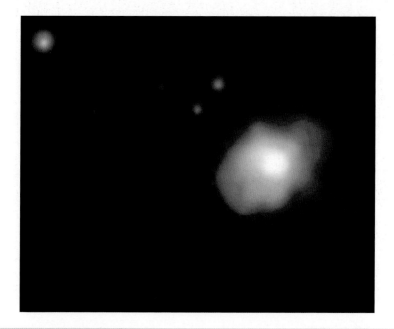

Fig. 8.14. The "Homunculus" around Eta Carina – Sky-Watcher 180 mm. Maksutov Cassegrain f/30, RGB. S. Massey.

Fig. 8.15. Messier 64 in Virgo – 25 cm f/4.7 Newtonian, only 50 × 2.56s (L). S. Massey.

Fig. 8.16. Central region of Messier 31 in Andromeda – Meade LXD75 SN8 at f/4, mosaic of four fields each 200 × 2.56s (L). Courtesy of Mark Garrett.

Fig. 8.17. NGC 3242 in Hydra – 25 cm f/10 Newtonian (2X Barlow), 150 × 2.56s (of each RGB). S. Massey.

Fig. 8.18. The "Hourglass" at the heart of Messier 8 in Sagittarius – Sky-Watcher 180 mm Maksutov Cassegrain f/15, 150 × 2.56s (of each RGB). S. Massey.

Fig. 8.19. NGC 5128 in Centaurus – 25 cm f/5 Newtonian, 400 × 2.56s (of each RGB) in semi-moonlit sky. S. Massey.

Fig. 8.20. NGC 2237/44/46 in Monoceros – 80 mm f/5.6 refractor. Mosaic made from nine separate fields, each 700 × 2.56s (L), *IR* block filter and 700 × 2.56s (of each RGB). Courtesy of Darrin Nitschke.

8.2 Dark Country Skies

(Figs. 8.21–8.42)

Fig. 8.21. Messier 83 in Hydra – 31.5 cm f/4.5 Newtonian, 500 × 2.56s (L), 100 × 2.56s (of each RGB). S. Quirk.

Fig. 8.22. The heart of Messier 20 in Sagittarius - 31.5 cm f/9 Newtonian, 100 × 2.56s (L), 100 × 2.56s (of each RGB) Various sense-up modes. S. Quirk.

Fig. 8.23. NGC 5128 in Centaurus – Vixen R150S 6" f5 reflector on an EQ6 Pro mount. L × 1464, R × 511, G × 521, B × 430. Courtesy of Darrin Nitschke

Fig. 8.24. NGC 1973–75–77 in Orion – 15 cm f/5 Newtonian, 700 × 2.56s frames, *IR* block filter. Courtesy of Darrin Nitschke.

Fig. 8.25. Messier 1 in Taurus - 25 cm f/5 Newtonian, 700 × 2.56s frames (L), *IR* block filter 230 × 2.56s (of each RB, sim G). Courtesy of Darrin Nitschke.

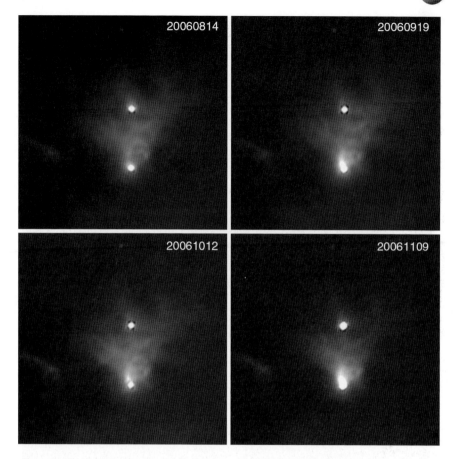

Fig. 8.26. NGC 6729 in Corona Australis – Shows patterns in the nebula changing as the star R CorAus varies in brightness - 31.5 cm f/4.5 Newtonian, each image 200 × 2.56s (L). S. Quirk.

Fig. 8.27. Barnard 33 and NGC 2023 in Orion - 10" f/6 Newtonian on Dobsonian mount. Courtesy of Martin Hilario.

Fig. 8.28. Barnard 86 in Sagittarius – 31.5 cm f/4.5 Newtonian, 200 × 2.56s (L), 100 × 2.56s (of each RGB). S. Quirk.

Fig. 8.29. Messier 104 in Virgo - GSO 10" f5 reflector mounted on an EQ6 Pro mount. L × 747, R × 228, G × 233, B × 254. Courtesy of Darrin Nitschke.

Fig. 8.30. NGC 2362 in Canis Major – 31.5 cm f/4.5 Newtonian, 200 × 2.56s (L), 100 × 2.56s (of each RGB). S. Quirk.

Fig. 8.31. NGC 7582–90–99 in Grus – 25 cm f/5 Newtonian, two fields merged, each 700 × 2.56s frames, *IR* block filter. Courtesy of Darrin Nitschke.

Fig. 8.32. NGC 104 in Tucana – 20 cm f/4 Newtonian and 0.6X focal reducer, 200 × 2.56s (L). S. Quirk

Fig. 8.33. NGC 1365 in Fornax – 25 cm f/5 Newtonian, 700 × 2.56s frames, *IR* block filter. Courtesy of Darrin Nitschke.

Fig. 8.34. NGC 5286 in Centaurus – 31.5 cm f/4.5 Newtonian, 100 × 2.56s (L), 100 × 2.56s (of each RGB). S. Quirk.

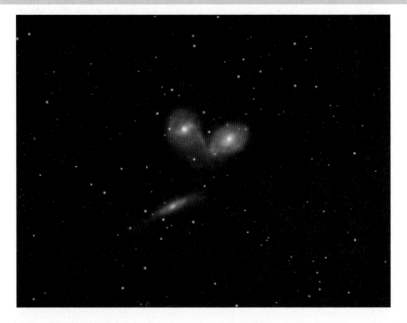

Fig. 8.35. NGC 6769–70–71 in Pavo – 25 cm f/5 Newtonian, 700 × 2.56s frames, *IR* block filter. Courtesy of Darrin Nitschke.

Fig. 8.36. Messier 27 in Vulpecula – 31.5 cm f/4.5 Newtonian, 400 × 2.56s (L), 100 × 2.56s (of each RGB). S. Quirk.

Fig. 8.37. Stephan's Quintet in Pegasus – 25 cm f/5 Newtonian, 700 × 2.56s frames, *IR* block filter. Courtesy of Darrin Nitschke.

Fig. 8.38. NGC 6744 in Pavo – 25 cm f/5 Newtonian, 700 × 2.56s frames, *IR* block filter, 230 × 2.56s (of each RGB). Courtesy of Darrin Nitschke.

Fig. 8.39. Shapley 1 in Norma – 31.5 cm f/4.5 Newtonian, 400 × 2.56s (L), 100 × 2.56s (of each RGB). S. Quirk.

Fig. 8.40. NGC 7331 in Pegasus – 25 cm f/5 Newtonian, 700 × 2.56s frames, *IR* block filter. Courtesy of Darrin Nitschke.

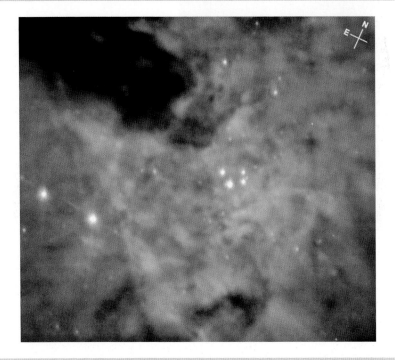

Fig. 8.41. The "Trapesium" at the center of Messier 42 in Orion – 31.5 cm f/9 Newtonian (2X Barlow), 125 × 2.56s (L), 100 × 2.56s (of each RGB). Various sense-up modes. S. Quirk.

Fig. 8.42. NGC 253 in Sculptor – 31.5 cm f/4.5 Newtonian, combination of two fields, each 400 × 2.56s (L), 100 × 2.56s (of each RGB). S. Quirk.

Glossary

Aberration
Deviation of light path from the ideal, caused by optical characteristics. Typically makes field of view edge appear out of focus and/or distorted. Includes spherical, chromatic, astigmatic, and coma-related defects.

Accumulation
Also see **Integrating**. In reference to cameras, capable of summing or co-adding a series of short-exposure images for output at longer preset intervals.

Array
Horizontal and vertical grid of pixels on a CCD image-sensing chip.

Astrometry
Positional measurement of objects on astronomical images.

AVI
Audio-video interleaved. Common video format of most commercial capture devices.

Blooming
Leakage of photoelectrons from a pixel that has exceeded its full-well capacity.

Binning
Summing the output of more than 1 pixel to increase sensitivity but at lower resolution. Typically small arrays of 2×2 or 3×3 pixels are used. Too big a sampling array will quickly over saturate the output signal.

CCD
Charge-coupled device. A solid state analog device consisting of an oxide silicon substrate and an array or a matrix of vertical and horizontal light-sensitive electrodes (detectors) on the surface called pixels (short for *picture elements*).

CCIR
Abbreviation for Comittee Consultatif International Radiotelecommunique (Fr), translates to International Radio Consultative Committee, a predecessor organization of the ITU-T. A consultative group governing recommended manufacturing standards for PAL signal cameras and encoding of interlaced analog video signals in digital form.

Chrominance
The color component of an image or a video signal.

CMYK
Short for Cyan, Magenta, Yellow, and Key (Black). In printing, the CMYK system works by partially or entirely masking certain colors on the typically white background (that is, absorbing particular wavelengths of light). Such a model is called subtractive because inks subtract brightness from white.

Collimation
The alignment of all optical components to produce an ideal and perfectly centered light path.

Cosmic ray
High-energy particle from outer space. Often seen interacting with the CCD chip on live video display.

Deep sky
Typically describes any astronomical object beyond our Solar System.

EIA
Abbreviation for Electronic Industries Alliance. Composed of trade associations for electronics manufacturers in the United States the group governs parts of the EIA standards, including the backward compatible extension of NTSC (RS-170).

EMI
Electromagnetic interference. A cause of unwanted noise being induced into poorly shielded signal cables from external electrical sources.

Flicker fusion
Eye-brain combination that fills in gaps in a video image stream, perceived as a fluid, continuous motion.

Full-well capacity
Total charge-holding capacity of a pixel. Also related to dynamic range.

Gain
Amplification of signal. Also increases the noise present in the signal.

Gamma
Supplementary brightness/contrast function for increasing mid-range brightness levels.

Histogram
Graph showing the values for brightness levels in an image.

Integration
See also **Accumulation**. Ability of a camera to collect or record photon build up in a CCD for a given length of time. Also, a descriptive name for image accumulation cameras.

Interlace
Alternating line output for a video signal. Analog video images often require the need for a deinterlacing filter to bring both video fields back into correct registration.

ISC
Image streaming cameras. These include CCTV cameras, camcorders, and webcams – any camera where rapid exposures are captured and output in rapid succession (resulting in the creation of a video movie).

JPEG
Joint Photographic Experts Group. Image file format with variable steps of compression. Standard for Web-published images.

Luminance
The brightness and contrast (greyscale) aspect of a video image.

Magnitude
Scale of brightness of a star or other object. A 1 magnitude difference is 2.512 times greater or less than the adjacent value.

MPEG
Moving Picture Experts Group. Type of compressed video file format.

NIT
Luminance value. One nit is equal to one candela per square meter (i.e., cd/m^2).

NTSC
National Television Standards Committee. An analog television standard used in Canada, the United States, Mexico, South Korea, Taiwan, and Japan, among others. It is a 525-line 30 fps television standard.

PAL
An abbreviation for phase alternating line. It is a color encoding system used in broadcast television in several countries around the world including the United Kingdom, Australia, and New Zealand. It is a 625-line 25 fps television standard.

Photometry
Measurement of the brightness of a star or an object in an astronomical image.

Pixel
Picture element. Photosensitive detector element in a CCD array.

Quantum efficiency
Percentage ratio of photons falling on a sensor to the electrical signal output.

RFI
Radio frequency interference. Another problem, like EMI, that creates unwanted signal noise in video images from radio sources.

Resampling
Changing the size of an image by interpolating pixel values.

Resolution
Amount of detail that an optical system or camera can reveal.

RGB
Red, green, and blue components of a color image. LRGB has an added luminance channel.

Seeing
Amount of turbulence in the atmosphere distorting light from its ideal path, consequently blurring detail in astronomical objects.

Sense-up
Term by the CCTV manufacturer Mintron in Taiwan for setting the accumulation mode or increased sensitivity of a video camera.

Signal-to-noise ratio (S/N)
Ratio of true signal to background noise in an image measured in decibels.

Spectrum
Range of wavelengths of electromagnetic radiation. Visible light ranges from about 400 to 700 nm. Without filtering, typical CCDs are sensitive to wavelengths reaching into the infrared at about 1,000 nm.

Stacking
Combining multiple images to average out background noise patterns, thus emphasizing "real" picture information with improved dynamic range.

TIFF
Tagged Image File Format. Also known as TIF. A popular high-color-depth image format.

Transparency (of the sky)
The "clearness" of the atmosphere to allow starlight to penetrate to the Earth's surface. Dust, water vapor, or smoke particles can absorb many wavelengths of light that dim astronomical objects.

Transparency image layers
Relates to the opacity between layered images in computer processing, revealing more or less of a background image.

Tri-color
Color image made from separate red-, green-, and blue-filtered black and white images.

Vignetting
Darkening at the edge of an optical path, caused by a constriction (such as the field stop of an eyepiece) of the light beam.

Wavelength (of light)
Distance from peak to peak of particular frequencies in the electromagnetic spectrum.

Wavelet
Mathematical manipulation function used in processing detail in an astronomical image.

For Reference and Further Reading

Berry, Richard, and James Burnell. *The Handbook of Astronomical Image Processing.* Richmond, VA: Willmann-Bell, Inc., 2005.

Buil, Christian. *CCD Astronomy: Construction and Use of an Astronomical CCD Camera.* Richmond, VA: Willmann-Bell, Inc., 1991.

Cooke, Antony. *Visual Astronomy Under Dark Skies.* NY: New York, Springer Publishing, 2005.

Horne, Johnny. "Videoing the Night Sky," *Sky and Telescope Magazine.* September/October 2007. p. 64.

Massey, Steve, and Steve Quirk. *Atlas of the Southern Night Sky.* Australia: New Holland Publishers, 2007.

Massey, Steve, Thomas A. Dobbins, and Eric J. Douglass. *Video Astronomy.* Cambridge, MA: Sky Publishing Corporation, 2nd edition, 2004.

Moore, Patrick. *Astronomy Encyclopedia.* London, UK: Octopus Publishing Group, 2002.

Quirk, Steve. "Enhanced Viewing and Imaging with the GSTAR-EX Camera," *Australian Sky and Telescope Magazine,* July/August 2007. p. 74.

Internet Sites of Interest

Adirondack Video Astronomy, (StellaCam video cameras), www.astrovid.com

AstroVideo, (Video software), www.coaa.co.uk/astrovideo.htm

K3CCDTools, (Video software), www.pk3.org/Astro/

Loreal, (Dark Ring removal freeware), http://astrosurf.com/hfosaf/

Lumicon, (Focal Reducers and adaptors), www.lumicon.com

MyAstroShop, (GSTAR-EX Deep sky and planetary CCD video camera), (GSTAR capture software), www.myastroshop.com.au

QCUIAG, (QuickCam and Unconventional Imaging Astronomy Group), www.qcuiag.co.uk

The Imaging Source, (USB and Firewire Progressive Scan cameras), www.theimagingsource.com

SCS Astro, (Webcam and Video Imaging Cameras), www.scsastro.co.uk

VideoAstro, (A Video Astronomy Discussion Group), tech.groups.yahoo.com/group/videoastro/

Index

Index

Index

Printed in the United States of America